Patrick Hamilton

The Resources of Arizona

Its mineral, farming, and grazing lands, towns, and mining camps, its

rivers, mountains, plains, and mesas, with a brief summary of its Indian

tribes, early history, ancient ruins, climate, etc.

Patrick Hamilton

The Resources of Arizona
Its mineral, farming, and grazing lands, towns, and mining camps, its rivers, mountains, plains, and mesas, with a brief summary of its Indian tribes, early history, ancient ruins, climate, etc.

ISBN/EAN: 9783337301132

Printed in Europe, USA, Canada, Australia, Japan

Cover: Foto ©berggeist007 / pixelio.de

More available books at **www.hansebooks.com**

THE

RESOURCES OF ARIZONA:

ITS

MINERAL, FARMING, AND GRAZING LANDS, TOWNS, AND MINING
CAMPS; ITS RIVERS, MOUNTAINS, PLAINS, AND MESAS;
WITH A BRIEF SUMMARY OF ITS INDIAN TRIBES,
EARLY HISTORY, ANCIENT RUINS,
CLIMATE, ETC., ETC.

A MANUAL OF RELIABLE INFORMATION CONCERNING THE TERRITORY.

COMPILED BY
PATRICK HAMILTON,
Under authority of the Legislature.

PRESCOTT, ARIZONA.
1881.

CONTENTS.

PREFACE.

To meet the growing demand for information concerning the Territory of Arizona, the Legislature, at the session of 1880–81, authorized the publication of this pamphlet. In the following pages the author has endeavored to present this information in such a compact and concise form as would more fully meet the many inquiries of people desirous of coming to the Territory, and at the same time convey to the general public a comprehensive idea of the country, and its vast and varied resources. The facts presented have been gained by personal observation, during a residence of several years in the Territory, and it is believed they constitute a full and impartial description of Arizona as it is to-day—its mineral, pastoral, and agricultural resources, towns, and settlements, with a glance at its past history and a few words regarding its future prospects. This being a publication authorized by the representatives of the people and paid for out of the public treasury, having no private scheme to advance, or no private interest to foster, the exact truth has been sought and the statements can be considered reliable. No portion of the Territory has been overlooked and no material interest has been neglected. While not claiming for this compilation exemption from all errors or mistakes, it is believed that such only will be found as are inseparable from a work of this nature. For valuable assistance in the collection of the data herein contained, the author is indebted to many gentlemen throughout the Territory, and takes this method of tendering his sincere thanks to one and all. With the hope that the publication may, in some measure, meet the want for authentic and reliable information about the "coming country," and help to convey to the outside world some adequate idea of Arizona and its grand resources, it is left with the reader to say how well the task has been performed.

PATRICK HAMILTON,
Commissioner.

AN ACT

TO AUTHORIZE THE PUBLICATION OF INFORMATION OF THE RESOURCES
OF ARIZONA TERRITORY.

*Be it enacted by the Legislative Assembly of the Territory of
Arizona:*

SECTION 1. That Patrick Hamilton is hereby constituted and
appointed a Commissioner to prepare, and cause to be published,
reliable information upon the mineral, pastoral, agricultural,
and other resources of the Territory; also, the cost and facilities of coming to the Territory, and such other general information as he may consider of value to capitalists desirous of investing in our mines, or to persons who may wish to immigrate to
the Territory.

SEC. 2. It shall be the duty of said Commissioner to collect
and prepare the information aforesaid by January 1, 1882, and
he is hereby authorized to contract for the publication of ten
thousand copies, in pamphlet form, upon the most reasonable
terms that the work can be done, provided that the cost of such
publication shall not exceed fifteen hundred dollars ($1,500).

SEC. 3. It shall be the duty of said Commissioner to distribute
said pamphlets in the cities and railroad centers of the Eastern
States, and on the Pacific coast, in such a manner as will give
them the widest and most useful circulation, and he shall furnish thirty copies to each member of the Eleventh Legislative
Assembly.

SEC. 4. It shall be the duty of the Territorial Auditor, upon
the completion of said publication, to examine the same, and if
found in accordance with the provisions of this act, he shall
give the said Commissioner a certificate, setting forth that the
work has been performed according to law.

SEC. 5. It shall be the duty of said Commissioner to keep a
correct account of the number of copies of said publication distributed by him, and to whom, and such other information in
connection therewith, as he may deem of interest, and to make
a full report of the same to the Governor of the Territory on or
before January 1, 1883, and the Governor shall transmit a copy
of said report to the next Legislative Assembly.

SEC. 6. Said Commissioner shall receive as compensation, for
the collection, preparation, and distribution of such information
the sum of two thousand dollars.

SEC. 7. Upon the completion of said publication, the Commissioner shall certify to the Territorial Auditor the amount due

for said work and to whom; and the Territorial Auditor shall draw his warrant for the amount in favor of the person to whom the same is due, as shall appear by the certificate of said Commissioner; and the Territorial Treasurer is hereby authorized and directed to pay said warrant out of any money in the Treasury not otherwise appropriated.

SEC. 8. This act shall take effect and be in force from and after its passage.

J. F. KNAPP,
Speaker of the House of Representatives.

· MURAT MASTERSON,
President of the Council.

Clause 12 of the Appropriation Act, passed subsequent to the foregoing, enacts as follows :

Twelfth. The sum of four thousand five hundred ($4,500) dollars is hereby appropriated to pay the Commissioner selected to compile, publish, and distribute the pamphlet on the "Resources of Arizona Territory," and the Territorial Auditor is hereby directed to draw his warrant on the Territorial Treasurer for the above amount, in favor of the Commissioner named in the act, and the Territorial Treasurer is hereby authorized and directed to pay said sum to said Commissioner out of any moneys in the Territorial Treasury not otherwise appropriated, in the manner provided for by the provision of said act.

THE RESOURCES OF ARIZONA.

HISTORICAL.

The region now embraced within the territory of Arizona, was first penetrated by Europeans nearly three hundred and fifty years ago. A quarter of a century before the founding of San Augustine, and long before Puritan or Cavalier had established themselves at Plymouth Rock or Jamestown, Spanish adventurers had explored the wilds of Arizona and New Mexico. Alvar Nunez de Vaca, one of the followers of Pamphilo de Narveaz, in his disastrous expedition to the coast of Florida, in 1538, being left by his commander, with four companions, on the desolate shore, resolved to penetrate the great unknown wilderness to the westward and join their countrymen in Mexiico. Without compass or provisions, they struck across the continent, discovered and crossed the Mississippi two years before De Soto stood upon its banks and found a burial place beneath its turbid waters. They traversed the great plains of the West, entered New Mexico, visited the pueblo towns, passed through the country of the Moquis, and, after many hardships and privations, joined their countrymen at Culiacan, in Sinaloa. They gave glowing accounts of the country through which they passed, and their description of the "Seven Cities of Cibola," the Moquis towns, excited the spirit of adventure and cupidity among the Spanish conquerers, and fired the zealous ardor of the missionaries. Padre Marco de Niza, under the patronage of the Viceroy Mendoza, set out from Culiacan in 1539, accompanied by a single companion, in search of the fabulous "Seven Cities." They passed through the Papagueria and the country of the Pimas, by the valley of the Santa Cruz and into the country of the friendly Yavapais, and at last came in sight of the goal of their arduous quest. Father de Niza sent his companion ahead, with some Indians, who had accompanied them from the Gila. The Moquis massacred the whole party. Father de Niza did not enter the city. He set up the cross, named the country the New Kingdom of San Francisco, and returned to Culiacan.

The public mind in New Spain was greatly excited by the news which the good father brought on his return. The thirst for gold and glory, and the desire to extend the influence of the cross, bore down all opposition. The Viceroy, Mendoza, projected two expeditions to explore the marvelous country to the north; one by land under Vasquez de Coronado, and the other by sea under Fernando Alarcon. In April, 1540, Coronado marched from Culiacan with nearly a thousand men (principally Indians). He visited the ruins of the Casa Grande, on the Gila, and in forty-five days after starting, reached the first of the "Seven Cities." Instead of the rich and populous region which their imagination had pictured, they found a poor and insignificant village. The province was composed of seven villages, the houses being small and built in terraces, as they are at the present day. The inhabitants were intelligent and industrious. They raised good crops of corn, beans, and pumpkins, dressed in cotton cloth, and were the same in all respects as their descendants, the Moquis and Zunis, are at the present time. Coronado penetrated to the New Mexican pueblos on the Rio Grande, explored the country as far east as the Canadian river, and north to the fortieth degree of latitude. Disappointed in his search for the riches he expected to find, the expedition returned to New Spain in the spring of 1542. The expedition of Alarcon sailed about the same time Coronado marched. The Gulf of California was discovered, and named the Sea of Cortez. The Colorado and the Gila rivers were also discovered. Two boats ascended the former stream to the Grand canyon. For forty years after these expeditions, no further efforts were made to explore the country. In 1582, Antonio de Espejo penetrated the country northward and discovered many populous pueblos in the Rio Grande valley, which are not mentioned by the historian of Coronado's expedition. He visited the Zunis, and passed westward to the Moquis, who met him with presents of corn and mantles of cotton. Forty-five leagues south-westward from the Moquis villages, he discovered rich silver ore in a mountain easily ascended. Numerous Indian pueblos were found in the vicinity of the mines, and two rivers, on which grew wild grapes, walnut trees and flax, were also discovered. Those streams were no doubt the Little Colorado and the Verde. More than a century elapsed after these explorations before any permanent settlement was made in the territory now known as Arizona. Towards the close of the seventeenth century, the Jesuit Fathers established the missions of Guevavi, Tumacacori, and San Xavier. Missions had been established some time before among the Moquis. In 1720, there were nine missions in a prosperous condition within what is now the territory of Arizona. The fruits of the untiring labors of the zealous fathers were shown in the peaceful and industrious Indian colonies which sprang up around their missions. Despite the expulsion of their founders, the Jesuits, in 1767, and the constant raids of the savage Apaches, the missions continued to flourish and grow rich, until the revolution for Mexican independence. Deprived of the protection of the vice-regal government, and constantly harassed by the Apaches, they languished and declined, until they were finally suppressed under a decree of the Mexican government in 1827. By the treaty of Guadalupe Hidalgo, in 1846, all that portion of the present Territory of Arizona north of the Gila river was ceded to the United States. At that time the population of the Territory was confined to a few hundred souls within the presidios of Tucson and Tubac. What is now known as northern and central Arizona did not contain a single white settlement. Outside the Pima and Maricopa villages on the Gila and Rio Salado, and the Moquis towns in the extreme north-east, the savage Apache was lord of mountain, valley, and mesa. In 1854, that portion of the Territory between the Gila river and the line of Sonora was acquired from the Mexican government by purchase. It was long known as the "Gadsden Purchase," the negotiations for its acquisition having been conducted by the Hon. James Gadsden, then minister to Mexico. The price paid was $10,000,000, and, in the light of its recent developments of marvelous mineral wealth, it can be considered a good bargain. Tubac and Tucson were taken possession of by the United States troops in 1855; the Mexican colors were lowered, the Stars and Stripes hoisted in their stead, and the authority of the United States established where Spaniard and Mexican had held sway for nearly 300 years.

Subsequent to the treaty of Guadalupe Hidalgo, the Territory formed a part of New Mexico. A memorial was presented to the Legislature of New Mexico on the first day of December, 1854, for a separate territorial organization. The name first adopted was "Pimeria;" but it was afterwards changed to "Arizona." The word Arizona is said to be derived from two Pima words: "Ari," a maiden, and "Zon," a valley, or country. It has reference to the traditionary maiden queen who once ruled over all the branches of the Pima race. Before the name was conferred on the whole Territory, it was borne by a mountain adjacent to the celebrated Planchas de Plata mines near the southern line of the Territory. Arizona remained a portion of New Mexico until the twenty-fourth of February, 1863, when the act was passed organizing it as a separate Territory. The civil officers appointed by the President entered the Territory on the twenty-seventh of December, 1863, and two days later, at Navajo Springs, the national colors were given to the breeze, and the Territorial Government formally inaugurated. The seat of government was established at Fort Whipple, in Chino valley, on the headwaters of the Verde. It was afterwards removed to Prescott, where it still remains.

The history of Arizona from the establishment of a Territorial organization up to the year 1874 has been a series of fierce and bloody struggles with the savage Apaches, and of slow but steady growth. The intrepidity, daring, and self-sacrifice of the early pioneers, who won this rich domain, foot by foot, from its savage occupants, yet remains to be written, and will be one of the bloodiest pages in the history of our frontier settlements. The hostile tribes were conquered and placed on reservations by General Crook, in 1874, and since that time the Territory has made rapid progress in population, wealth, and general development. With the opening of a transcontinental railroad across the southern portion of the Territory, and the discovery of immense veins of silver ore adjacent thereto, Arizona has attracted the attention of the whole country, and capital and emigration have flowed in upon her at an unexampled rate. One of the first-discovered portions of North America, so long neglected and unknown, is at least beginning to yield up those treasures which for ages have remained hidden in its mountain fastnesses, guarded by the fiercest of savages. A new era has dawned for Arizona—an era of peace, progress, and prosperity. The demon of isolation and the curse of savage dominion, which so long brooded over the land, have been swept aside by the advancing tide of civilization, and Arizona's future is bright with the promise of a powerful and a prosperous state.

GENERAL DESCRIPTION OF ARIZONA.

The Territory of Arizona comprises the extreme south-western portion of the United States. It is bounded on the north by Nevada and Utah, on the east by New Mexico, on the south by Sonora, on the west by California and Nevada. It extends from the one hundred and ninth meridian west to the Great Colorado; and from ·31° 28′ of north latitude to the thirty-seventh parallel, and contains an area of about 114,000 square miles. The physical features of the Territory may be described as a series of elevated plateaus, having an altitude of from 100 feet in the south-west, up to 6,000 and 7,000 feet above the sea level, in the north. Mountain ranges, having a general direction of north-west by south-east, extend over this lofty plateau the entire length of the Territory. These mountains often present the appearance of broken and detached spurs, and sometimes occur in regular and continuous ranges. Narrow valleys and wide, open plains lie between the mountains, while deep canyons and gorges, formed by the rains and floods, which sometimes rush with irresistible force from the mountain barriers, cross the country in every direction. The most extensive of these grand mesas, or table lands, is the Colorado plateau, in the northern portion of the Territory, occupying nearly two-fifths of its entire area. This great plateau has an average altitude of between 5,000 and 6,000 feet. Its surface is diversified by lofty peaks and isolated ranges; it is covered nearly its entire extent with fine grasses; it is penetrated on the west by the Rio Colorado, which has worn a channel thousands of feet in depth. It is also cut by the San Juan on the north-east, and the Little Colorado, the Verde, the Salinas, and the San Francisco on the south. These rivers form in places deep gorges, and again widen into beautiful and productive valleys. Perhaps nowhere on the continent can be found a more striking panorama of mountain, valley, mesa, and canyon. From north to south, from east to west, the country is crossed by mountain ranges and isolated peaks of strange and fantastic shapes. In the eastern portion of the Territory, extending from the San Francisco Mountain on the north, to the Gila river on the south, a long line of extinct volcanoes can be traced, and immense lava fields, which are found in different portions of the Territory, prove conclusively that Arizona was, in ages past, the scene of active eruptive agencies.

The south-western portion of the Territory may be described as composed of wide plains, covered in places with a sparse growth of grass, and dotted with peaks and detached spurs. The south-eastern portion of Arizona is made up of mountain ranges, which sometimes rise into commanding peaks, like the Santa Ritas and Mount Turnbull, with grassy plains and rich valleys lying between. The central portion of the Territory can show some of the most attractive scenery on the continent. It is also well watered, and contains the largest body of agricultural land in Arizona—the valleys of the Gila and Salt rivers. One of the wonderful curiosities of the Territory is the Grand canyon of the Colorado. This is one of the most stupendous chasms to be found on the continent, and probably has not its equal on the globe. It is a tremendous gorge, 400 miles in length, and from 1,500 to 6,000 feet in depth, cut through the eruptive rock by the river, in its passage for ages from its mountain sources to the sea. Down in the gloomy recesses of this forbidding gorge, which calls to mind the portal to Dante's Inferno, the light of day hardly ever penetrates, and the river, looking like a slender silver thread, foams and whirls among the rocks and falls which impede its progress. The canyon was first discovered by Coronado's expedition in 1540, and its length and depth accurately measured. It has been explored its entire length by Major Powell, who has given a most interesting and vivid description of its many wonders. The Little Colorado, one of the main tributaries of the great river, has also a canyon system of its own, but on a much smaller scale than the larger river.

The geological character of the Territory exhibits almost every formation to be found on the continent. North of the Grand Colorado and the Colorado Chiquito, the surface rock is a pure sandstone. The main ranges through the central portion of the Territory are composed of granite, porphyry, and slates. The mountains extending south-east from the great cone of the San Francisco to the thirty-fourth parallel, are mostly of volcanic origin. Between the Gila and the Sonora line is found granite, limestone, porphyry, trap, and much metamorphic rock. The lower portion of the Great Colorado basin bears traces of violent volcanic disturbance, and is covered in places with scoria and ashes; its upper portion is composed of granites, porphyry, and slates, with here and there isolated ranges and jagged peaks scorched and riven by the fiery flood which has swept over this part of Arizona in ages when our earth was young.

Arizona is a land of marvels for the scientist and the sight-seer. Nowhere on the globe can the operations of nature be traced more clearly and distinctly. Torn and riven by stupendous gorges and deep canyons, crowned by lofty mountains, and diversified by immense plains, grassy parks, beautiful valleys, and elevated mesas, the topography of the country in variety, weird beauty, and massive grandeur, is not excelled on the continent. That the great plateau of Arizona was once an inland sea, there can be little doubt; and the isolated mountain masses, rising like islands above its surface, and the fantastically castellated buttes, which dot its immense plains, show clearly the erosion caused by the retreating waters. Arizona is a land that offers to the geologist and mineralogist a field both interesting and instructive; a land where the great book of nature lies open, with the record of countless ages stamped on its broad pages.

RIVERS AND MOUNTAINS.

RIVERS.

First among the rivers of Arizona is the Colorado, which washes the western border of the Territory. This mighty stream is the principal tributary of the Pacific ocean, on the North American continent, south of the Columbia. It belongs to that vast system of water-courses which have their sources in the Rocky mountain cordilleras, and drain the continent from ocean to ocean. The great river was discovered on the ninth of May, 1540, by Captain Fernando Alarcon. He ascended the stream in boats 85 leagues from its mouth. He also discovered the Gila and called it the Miraflores. The Colorado takes its rise in the Wind River chain of the Rocky mountains, in latitude 43° 30′ north, and some 12,000 feet above the level of the sea. It flows towards the south-east in its upper course, and is called Green river. Below its junction with the Grand, its great tributary from the east, its course is south-westerly until it is joined by the San Juan, above the entrance to the Great canyon. From there it runs south-westerly through the great chasm of the Colorado plateau to the mouth of the Virgin, and from that point to the Gulf of California it winds its way almost due south. The length of the Colorado and its tributaries is nearly 2,000 miles, draining an area larger than New England, Pennsylvania, and Virginia combined. Above its junction with the Grand its waters are clear and limpid, but after passing through the Great canyon they assume a reddish hue, and are as turbid as those of the Missouri. The river is navigable for over 600 miles by boats of light draught, but the constant changing of its channel makes navigation difficult and sometimes hazardous. The valley of the Colorado below the canyon, although narrow, for such a stream, and subject to overflow, contains hundreds of thousands of acres of productive soil. The great Colorado drains the entire Territory of Arizona, and every stream and water-course within its borders finds its way to the mighty river. The Gila river, the next in size to the Colorado, takes its rise among the Mogollon mountains in New Mexico, on the divide that separates the waters of the Rio Grande and those flowing into the Gulf of Mexico, from those that flow westward to the Pacific. In the upper part of its course, the Gila is a mountain stream, dashing through rocky gorges, deep canyons, amid the wildest of mountain scenery. It forms no valley of any size, except the site of old Fort West, until it passes the one hundred and ninth meridian of longitude and enters the Territory of Arizona. A few miles west of the boundary line it receives the San Francisco from the north, a mountain stream bordered by a narrow valley. Some miles below it is joined by the waters of the Rio Prieta. At this point the valley of the Gila rapidly widens into a rich and productive stretch of bottom land, known as Pueblo Viejo, which extends west for nearly 50 miles. The Rio Bonito, a clear mountain stream, enters the Gila from the north, near the head of this valley. The San Carlos, which rises in the White mountains, joins the Gila at the lower end of the Pueblo Viejo valley. It is a fine mountain stream, with a rich and beautiful valley, now occupied as an Indian reservation. Below the San Carlos, the Gila flows through a deep and rocky canyon formed by the Mazatzal, Mogollon, and Mescal ranges from the north, and the Galiuro and Pinaleno ranges from the south. Just below the canyon the San Pedro unites with the Gila, from the south. The San Pedro is a narrow, swift stream, about 25 feet in width, and averaging about three feet in depth. It rises in Sonora and flows north through a fertile valley, with grass-covered mesas on each side, which sometimes rise into lofty ridges. Its course through Arizona is about 100 miles. The Arivaypa joins the San Pedro one mile below old Fort Breckenridge. It is a clear, beautiful stream, flowing through a rocky gorge, with a narrow valley of great fertility.

Below the canyon, the Gila forms a valley from one to five miles wide, which produces luxuriant crops by irrigation. The Santa Cruz, from its source in the Huachuca mountains, after flowing southward through Sonora, making a curve to the west, and passing by Tubac and Tucson, enters the Gila by an underground channel below the Pima villages. Salt river unites with the Gila at the point of the Sierra de Estrella. It is a bold and rapid stream, having its source in the White mountains, and carrying a volume of water nearly three times as large as that of the Gila. Its upper course is through deep canyons, occasionally widening into narrow and fertile valleys. The main branch of the stream is known as Black river, flowing through a rugged, mountainous country. It receives the White Mountain, Carizo creek, Tonto creek, and other streams from the north, above the canyon, and the Rio Verde below that point. After breaking from the canyon the Salt river runs in a south-westerly direction, through a wide plain, containing the largest body of farming land in the Territory. The river is fed by mountain springs and snows, and carries a large volume of water. It is about 150 miles in length.

The Rio Verde rises in Chino valley, in the great plateau that stretches south from the San Francisco mountains, near latitude 35° 30′ north. It pursues a southerly direction, most of the way through a beautiful and productive valley, receiving in its course Oak, Beaver, and Clear creeks from the east, and Granite creek from the west. It joins the Salt river a few miles below Fort McDowell. The length of the Verde is nearly 150 miles. It carries a volume of water almost equal to the Gila, and is one of the finest streams in the Territory. The Hassayampa and the Agua Fria take their rise in the Sierra Prieta, near Prescott, and enter the Gila below the Big Bend, but they sink in the thirsty sands long before they reach that stream. The Gila and its tributaries drain more than one half of the Territory. The river is about 500 miles in length, four-fifths of the distance being through Arizona.

The Colorado Chiquito takes its rise in the Sierra Blanco, near the line of 34° north. The country around its headwaters is covered with pine forests and dotted with beautiful mountain lakes. It pursues a north-westerly direction, and enters the Great Colorado, through a canyon half a mile in depth, 200 miles from its source. During its journey it is joined by the Rio Puerco and the Zuni river, from the north, and by Silver and Carisso creeks, and other inconsiderable streams, from the south. The upper valley of the Little Colorado is rich and fertile, producing fine crops with irrigation. Williams Fork empties into the Colorado on the line of 34° 20′ north latitude and 114° 8′ west longitude. The Santa Maria, the eastern branch of this stream, has its rise in the Juniper range, north-west of Prescott, while another branch rises at Peeple's valley. They join the Big Sandy, that has its source in the Cactus pass, and thence flow westward to the Great river. These are the important water-courses of the Territory, though there are many others which in rainy seasons pour their turbid floods into the Colorado and the Gila.

MOUNTAINS.

The mountains of Arizona are among the most interesting physical features of this wonderful country, and would require a volume to describe them in detail. It can be said that they show very little regularity, although they have a marked parallelism in the trend and direction of their axis, from north-west to south-east. The parallel ridges of the Great plateau diverge from two points within the limits of the Territory—the Great canyon of the Colorado, and the canyon of the Gila above the junction of the San Pedro. Beginning 40 miles south of the Little Colorado, the San Francisco peak, the highest in the Territory, rears its lofty head nearly 12,500 feet above the level of the sea. The San Francisco may be considered the northern point of the great ranges which extend from the one hundred and ninth to the one hundred and thirteenth degree of longitude, and from the thirty-sixth parallel to the Sonora line. That part of this range north of the Gila canyon is known as the Mazatzal, and farther east as the Mogollon and Sierra Blanco. There are also many detached spurs and isolated ranges, such as the Superstition, the Mescal, the Apache, the Pinal, the Gila range, and the Sierra Natanes. Most of these mountain ranges are covered with timber, and the Mogollon, Mazatzal, Sierra Blanco, and Pinal mountains, have a splendid growth of pines, cedars, oak, and juniper. Rich and nutritious grasses cover these mountains and the high table lands adjacent. Many of the ranges attain lofty elevations. The Sierra Blanco is 11,-300, and the Four Peaks, in the Mazatzal range, nearly 10,000 feet above the level of the sea. South of the canyon of the Gila, on the line of 32° 30', the parallel ranges of the Great plateau system are clearly defined, all having the north-west and south-east trend.

The Peloncillo, the Pinaleno, the Galiuro, the Chiricahua, the Santa Catarina, the Huachuca, the Santa Rita, the Dragoon, and Whetstone are the most prominent. Nearly all of these mountains are well watered, and covered with grass and timber. Mount Graham, in the Pinaleno range, attains a height of 10,-500 feet above sea level, while the lofty peak of Mount Wrightson, in the Santa Ritas, has exactly the same elevation. West of Tucson, in the Papago country, are several isolated ranges, of which the highest is Baboquivara peak, standing like a giant sentinel, guarding the weird fastnesses of the Papagueria. The Arizona mountains, which have given their name to the Territory, extend from the point of the mountain, north of Tucson, into Sonora. They are sometimes called the Tubac mountains, and the Atascoso. They are of volcanic origin, broken and irregular. North of the thirty-fourth parallel, and west of the Verde, is the ridge that separates the waters of the Rio Verde from the Agua Fria, known as the Verde mountains.

West of the range are the Bradshaw and Sierra Prieta, that girdle Prescott, and, extending north, join the Santa Maria and Juniper mountains. The Bradshaw and Sierra Prieta are massive ranges, well watered and thickly covered with pine, oak, and juniper timber, with a fine growth of grasses. Mount Union, in the Sierra Prieta, nine miles south of Prescott, attains an elevation of 9,000 feet. In the basin of the Colorado, the principal ranges are the Sacramento, the Cerbat, the Hualapai, the Peacock, the Cottonwood, and the weird and desolate Music mountain, in the north; and in the south, the Harcurar, the Plomosa, the Castle Dome, and the Chocolate ranges. Most of these run parallel to the course of the Great river, with immense open valleys between. They are generally devoid of timber, and many of them bear the marks of violent volcanic action. There are many other detached ranges, such as the Black Hills, east of Prescott, rich in mineral and covered with timber; the Antelope, west of the Bradshaw, famous for its gold placers, and Bill Williams mountain, south-west from the San Francisco peak. The mountain system of Arizona partakes of the character of the Sierra Nevadas and the Rocky mountain cordilleras, and may be considered outlying spurs from both. In 43° 30' the Wind river chain of the Rocky mountains divides around the sources of the Colorado. One branch trends to the south in the Great Wasatch range, and, widening out to the level of the Great plateau, reaches the Grand canyon of Colorado on the line of the one hundred and twelfth degree of longitude. A branch of the Sierra Nevada leaves that range south-west of Owens river, and, with a general sweep to the south-west, merges into the plateau and joins the Wasatch at the Great canyon. Our space will not permit us to give more than a passing glance at the grand system of sierras, crowned with their lordly pines, and holding in their rocky embrace vast mineral wealth, which are such a magnificent feature of the topography of Arizona.

FAUNA AND FLORA.

FAUNA.

The fauna of Arizona, in its extent and variety, will compare with any portion of the Union. Nearly all the animals indigenous to the temperate zone are found throughout the Territory, and in some localities it is the very paradise of the sportsman. The grizzly bear is found in the White mountain range, near Camp Apache; the cinnamon and the black bear are met with in the San Francisco, the Mogollon, the Sierra Blanco, the Bradshaw, the Mazatzal, the Chiricahua, the Huachuca, the Santa Rita, and in nearly all the wooded mountains of the Territory. The coyote, or prairie wolf, roams through the length and breadth of Arizona. The black-tailed deer is common in the northern and central portions of the Territory; it attains a large size, and some weighing 250 pounds have been killed. The California lion, or cougar, makes his home in every county in the Territory. The antelope is found in large bands on the elevated mesas and grassy plains that stretch from the Patagonia mountains to the Cocouino forest; the big-horn mountain sheep is a dweller in the almost inaccessible crags and barren mountain peaks of northern Arizona. Although the elk can hardly be considered a native of this latitude, some large specimens have been seen in the lofty ranges of the San Francisco and the Sierra Blanco. The fox and the wildcat are extensively distributed, some of the latter reaching a very large size. The wood rat, the kangaroo rat, and the white mouse, are found in all parts of the Territory; gophers are numerous, the black-faced variety being mostly confined to the Sierra Blanco; squirrels are seen everywhere. The beaver inhabits the streams throughout the Mogollon, the White mountains, the Verde and its tributaries, and the San Pedro. Rabbits are found in every section. Arizona offers a fine field for the ornithologist; it is exceedingly rich in the number and variety of the feathered tribe. The wild turkey is found in the Bill Williams, San Francisco, Mogollon, Sierra Blanco, Chiricahua, on the headwaters of the Gila and Santa Cruz, and in nearly all the wooded mountains of southern Arizona. Wild duck are plentiful in the watercourses of northern, eastern and south-eastern Arizona, and the wild goose is occasionally seen on the Colorado, the Gila and the Salt rivers. The crested quail, or California partridge, is extensively distributed and rapidly increasing since the settlement of the country by whites; doves and pigeons are found in the mountains and elevated plateaus. The western hawk inhabits all parts of the Territory. The crow family is well represented and is met in every direction. The American eagle is found among the lofty peaks and deep canyons of the Sierra Blanco. There are many species of the owl family, and their solemn hooting makes night hideous from the Utah line to the frontier of Sonora. The melody of the mocking-bird is heard in Arizona wherever there is a stream or a grove; sparrows abound in the southern and central portions of the Territory, and the sweet song of the thrush trills forth in many parts of eastern Arizona. The oriole is found in the region of Camp Grant; and humming-birds, warblers, and finches are met with in the central and south-eastern valleys and mountain ranges. Water-ousels and bluebirds frequent the elevated regions of the south-east. The Arizona vireo is one of our sweetest singers, and is widely distributed; wrens are numerous in the south; swallows, buntings, jays, grosbeaks, and many species of the woodpecker are found in every part of the Territory; blackbirds are at home everywhere. Such aquatic birds as herons, snipes, sandpipers, cranes, etc., are found along the Colorado, the Gila, the Salt, and the larger streams. To describe fully the birds of Arizona, would require a volume; in brilliancy of plumage, sweetness of song, and variety of species, the feathered warblers of the eastern portion of the Territory are not excelled in the Union.

FLORA.

The flora of Arizona has many distinct peculiarities, and embraces several varieties found nowhere else in the United States. For the botanist, the Territory presents a wide field for study and investigation. Arizona is the home of the giant cactus, called by the aborigines, the *sahuaro*. This plant sometimes reaches a diameter of two feet, and frequently attains a height of forty feet. Its body is pale green, fluted like a Corinthian column; gigantic arms, like the branches of a candelabrum, put out from the main trunk towards its top, the whole being covered with sharp, prickly thorns. The plant bears a purple blossom, and in the latter part of June a palatable pear-shaped

2

fruit, prized by Mexicans and Indians, and tasting something like a fig.

The frame of the *sahuaro* is composed of narrow sticks of wood, arranged in the form of a cylinder, and held together by the outside fibers. When this "giant of the plains" falls, these ribs of wood are used for roofing adobe houses, fencing, etc. The prickly pear, another species of the cactus family, is found on the elevated mesas throughout the Territory. It attains a height of from four to six feet; has large fleshy leaves, which, in their tender state, are cooked by the natives, and taste not unlike string beans. It bears a pink-colored, pear-shaped fruit, palatable and refreshing to the thirsty traveler. The vinegar cactus, another variety of the plant, bears a small, deep-red berry, exceedingly acid in taste, which is used by the Indians as an antiscorbutic.

The *bisnaga*, or "well of the desert," is one of the most valued varieties of the cactus; it seldom reaches a height of over four feet, is of a cylindrical shape, covered by sharp thorns. The plant grows on the foothills and elevated plains. By cutting out the center, a bowl-shaped cavity is formed, which soon fills with water, affording to the thirsty wanderer a refreshing drink; the *bisnaga* also bears a bright yellow fruit, which is not unpalatable. There are many other varities of the cactus in all parts of the Territory, one of the most uninviting being the *cholla*, which sometimes grows to a height of five feet, with numerous branches covered with bunches of coarse thorns. A beautiful plant, which in the spring puts forth green leaves and scarlet blossoms and is found all over the table lands of Arizona, is the *ocotillo*. It is by some classed with the cactus family; grows in clusters of straight poles, from ten to fifteen feet in height, covered with sharp thorns. The plant is used extensively for fencing in portions of the Territory where there is a scarcity of wood.

The *maguey*, or *mescal*, sometimes called the century plant, is found on every hill and plateau of Arizona, and is the most useful of all the natural vegetable products of the Territory. It is brought under a high state of cultivation in Mexico, and is a source of large revenue in many portions of that country. Its long, sharp-pointed green leaves branch from the root to a height of three or four feet; they are fleshy and stiff, their edges being covered with thorns. The center of the plant is a large head, something like a cabbage, from which springs a slender pole, eight to twelve feet in height, bearing near its top, short branches which produce a yellow flower. The head is the valuable part, and is looked upon by the Apaches as their chief article of food. In preparing it for use the leaves are peeled off, the head is placed in a primitive oven made of round stones sunk in the ground, and roasted; it is then ready for use, is sweet and nutritious, tasting like a boiled beet. The Indians also make it into flat cakes, which were their principal means of subsistence when on the war-path, during the long and bloody struggle against the whites. The juice is sometimes extracted, and makes a syrup that is very palatable; the Indians also ferment it and produce an intoxicating liquor called *tizwin*. The Mexicans distill the plant and make " mescal." It is as clear as gin, has the strong smoky taste of Scotch whisky, and will intoxicate as quickly as either. The Indians make ropes from the fiber of the plant, and a fine quality of paper has also been manufactured from it. Of all the plants growing within the limits of the Territory, it is the most valuable; it contains a large amount of saccharine matter, while its fibers can be utilized for the making of many useful articles.

The *amole*, or soap weed, is another of the valuable plants indigenous to Arizona, and grows all over its table lands and grassy plains. It reaches a height of three or four feet, with long and narrow pointed leaves, which make excellent ropes, paper, cloth, and other fabrics; the roots are used by the natives as a substitute for soap. For washing woolen goods it is superior to the soaps of commerce, the flannels being thoroughly cleansed without shrinkage; the roots are also used as a hair wash, keeping it soft and glossy.

The *hedeundilla*, or grease wood, covers the hills, table lands and dry plains of Arizona, over its entire extent. It grows from two to eight feet high, and in the early summer produces a yellow blossom; when the leaves are rubbed between the hands an unpleasant odor is produced and a greasy substance adheres to the fingers. A gum is obtained from this plant which is said to be valuable for medicinal purposes. Among the other useful plants of the Territory may be mentioned the *pectis* and the creosote bush; the former has an odor like essence of lemon. No doubt there are many other plants and shrubs, rich in medicinal qualities, which will be brought to light when the flora of Arizona is fully classified and thoroughly known.

Grapes, wild cherries, currants, strawberries, and blackberries, are found in the mountains and valleys of northern, central, and eastern Arizona. The native grapes are rich in saccharine matter, and produce a very palatable wine, tasting like light claret. Walnuts are plentiful in the mountains and foothills of central Arizona. The wild coffee plant is found on the plateau of the central portion of the Territory; the berry looks like the coffee of commerce and the flavor bears a slight resemblance to the domestic article.

Pine, cedar, and juniper, cover the mountains and table lands of northern and eastern Arizona; the great forest of the Mogollon range extends south almost to the Gila river and contains some of the finest pine timber on the continent. In the mountains south of the Gila, pine is found around the summits of the Pinaleno, the Santa Catarina, the Santa Rita, the Huachuca, and the Chiricahua ranges, while the rolling foothills are covered with a magnificent growth of oak. Sycamore, ash, walnut, elder, and cottonwood are found along the water-courses in all parts of the Territory. Among the valuable woods of Arizona is the mesquite. · This tree is a native of the region south of the Great plateau, and is nearly always found in good soil. Along the Gila, the Salt river, the Lower Colorado and the Santa Cruz valleys, large forests are often met with, many trees growing to a height of thirty feet. It is a close-grained wood, makes excellent wagon timber and splendid fuel. The tree is bushy in appearance, with a leaf resembling the locust; it bears large quantities of a bean-like fruit, which constitutes one of the chief articles of food among the Indians in the southern portion of the Territory. These beans make excellent food for cattle and horses. A dark-looking liquid exudes from the tree during the summer months, in color and consistency like gum arabic. The mesquite makes a handsome shade tree, and is one of the most valuable of the native woods of the Territory.

The *palo verde*, or green tree, is a native of the soil; it is found on the dry mesas, rolling hills and barren plains of the south and south-west. It seldom attains a height of over twelve feet; its branches are covered with thorns; its wood is soft and spongy, and it does not make even good fire-wood. The iron wood is a species of the mesquite, which it much resembles. It is a heavy, close-grained wood, susceptible of a high polish; when dry it is hard and brittle and almost impossible to cut with an ax. The bean it bears is similar to the mesquite, and contains as high as 35 per cent. of grape sugar; the Indians prize it highly as an article of food.

Of the grasses of Arizona, the most widely distributed is the black and white *gramma*, which grows in nearly every part of the Territory. A coarse grass called the *gayette* is found in the west and south-western portions of the country. In the higher regions, the pine, the mesquite, and other varieties, are met with. The coarse bunch grass, known as the buffalo, is found growing in many of the southern valleys and foothills. The grasses of the country are rich and nutritious, keeping stock in prime condition all seasons of the year.

FISH.

Although not coming properly within the scope of this division, something about the fishes of the Territory may not be out of place here. In the Colorado there is a species of the finny tribe known as the " Colorado salmon." They are a fish tasting something like a sturgeon, and reaching a large size, some weighing 70 pounds having been taken near Yuma. In the Gila there is a fish resembling a sucker; it is found in large numbers and is well-flavored. What is known as the " Verde trout " is found in that stream and its tributaries; it resembles the mountain trout, and were it not for the number of bones, would be a valuable food fish. A fish called the humpback is found in the Salt river, and some weighing four pounds have been taken. A fish resembling a trout is also found in the Salt; it is of little value, being composed mostly of bones. In the headwaters of the Colorado Chiquito, and in the cold and sparkling streams which flow down from the Mogollon and the Sierra Blanco mountains, trout are found in abundance. These streams, fed by heavy winter snows, are alive with this valuable fish, many of them weighing three and four pounds. In the upper waters of the Gila is found what is known as the white trout; it is a well-flavored and palatable fish. The Legislature of 1880–81 passed an act for stocking the rivers and lakes of the Territory with carp and other varieties suited to the climate. Already steps have been taken by the Commissioners appointed under the provisions of the act to carry out its objects, and Arizonans can hope in a few years to see the water-courses throughout the Territory well provided with a good supply of food fishes.

Newspaper correspondents and " tender-foot " visitors have given Arizona an unenviable notoriety for the number, size, and venom of its reptiles and poisonous insects. The truth is they are not as numerous or dangerous as in many of the Western States. On the rolling plains and mesas several species of the rattlesnake are met with, but are far less numerous than has been popularly supposed. It is stated that in one exploring expedition of over 2,000 miles, not more than twenty snakes were observed. In the lofty mountain ranges they are rarely met with. Small-sized lizards are found everywhere on the dry mesas and plains, and the horned toad is at home in many localities. The saurian known as the " Gila monster," is found in the southern part of the Territory; it is a large species of the lizard, and makes its home on the barren plains that stretch along the Gila and its tributaries, below the canyon. It is red and black in color, is covered with scales like the alligator, and is entirely harmless. It sometimes attains a length of two feet. This variety of the lizard is peculiar to Arizona, and is considered one of the institutions of this peculiar country.

COUNTIES AND BOUNDARIES.

The Territory of Arizona is divided into ten counties, namely: Pima, Yavapai, Maricopa, Mohave, Apache, Yuma, Pinal, Cachise, Gila, and Graham.

PIMA.

Pima county, the oldest inhabited portion of the Territory, is bounded on the north by Maricopa and Pinal, on the east by Cachise, on the south by Sonora, and on the west by Yuma county. The western portion of the county consists of dry, rolling plains, with isolated peaks and detached mountain ranges. It is covered with a sparse growth of grass, and in places, with mesquite wood. Water is scarce in this region, but wherever it is found grazing is excellent. Its mountains are rich in gold, silver, and copper. This part of the Territory is the home of the Papago Indians, and is known as the Papagueria. Pima county, south and east of Tucson, may be described as a country of plains, rolling hills, and lofty mountains. The Santa Ritas and the Patagonia ranges are well watered and timbered, while their slopes are covered with fine grasses. To the north the rocky Santa Catarina stretches away toward the canyon of the Gila. The Santa Cruz flows through the county, leaving a rich and productive valley. Pima has fine ' grazing lands, and its mountains are rich in minerals.

YAVAPAI.

Yavapai county extends from the thirty-fourth to the thirty-seventh degree of latitude, and embraces nearly three degrees of longitude. It contains nearly one-third of the entire area of the Territory. It is bounded on the north by Utah, on the east by Apache county, on the south by Maricopa and Gila counties, and on the west by Mohave county. It embraces the larger portion of the Great Colorado plateau, and its general elevation is from four to seven thousand feet above the level of the sea. Its physical features may be described as an immense elevated table land, crossed in all directions by lofty mountain ranges, adorned by beautiful valleys, and seamed and riven by deep canyons and rocky gorges. The mountains carry a fine growth of pine, oak, and juniper, while the uplands are covered with a luxuriant growth of nutritious grasses. The county is watered by the Colorado Chiquito, the Verde, the Agua Fria, the Hassayampa, the Santa Maria, and many other streams. That portion of the county south of the thirty-fifth parallel is rich in minerals of almost every description. The grazing resources of Yavapai are not excelled in the Territory. In the north-eastern corner of the county is that remarkable region known as the Painted desert, composed of mighty columns which have been left standing in solitary grandeur by slow denudations which have been at work for ages. This wild and weird region partakes of the character of the "Fata Morgana." Explorers say that on its air are depicted "palaces, hanging gardens, colonnades, temples, fountains, lakes, fortifications with flags flying on their ramparts, landscapes, woods, groves, orchards, meadows, and companies of men and women, herds of cattle, deer, antelope, etc., all painted with such an admirable mixture of light and shade that it is impossible to form any conception of the picture without seeing it." The Indians call it the country of departed spirits.

MARICOPA.

Maricopa county is bounded on the north by Yavapai, on the east by Gila and Pinal, on the south by Pinal and Pima, and on the west by Yuma. The western portion of the county is composed of broad plains, crossed by rugged mountains, covered with coarse grasses, with mesquite and palo verde wood growing in many places. The Gila river enters the county near Maricopa Wells and flows for nearly 100 miles through the western portion of it, making a rich and productive valley. Salt river, in its course through Maricopa, flows through the finest body of agricultural land in the Territory. That portion of Maricopa north and east of Phœnix, is a rugged, mountainous region, intersected by spurs from the Mazatzal and the Verde ranges, and known to be rich in minerals. Maricopa, besides its great agricultural and mineral resources, contains some fine grazing lands along the Gila, the Upper Salt, and the Verde.

MOHAVE.

Mohave county occupies the north-western corner of the Territory. It is bounded on the west by the Colorado river, on the north by Utah and Nevada, on the east by Yavapai, and on the south by Yuma. Mohave is a region of rugged mountain ranges, with immense valleys, covered with coarse but nutritious grasses. Four well-defined ranges, the Sacramento, the Cerbat, the Hualapai, and the Cottonwood, pursue a parallel course through that portion of the county south and east of the Colorado. Water is found in these mountains, and nearly all of them are mineral-bearing. Mohave has some fine grazing land, but its agricultural resources are limited to the valley of the Big Sandy and the Colorado. But little is known of the region north of the Colorado, though it is supposed to be an elevated plateau, crossed by mountains, seamed by canyons, and generally destitute of water.

APACHE.

Apache county occupies the north-eastern portion of the Territory. It is bounded on the north by Colorado, on the east by New Mexico, on the south by Graham and Gila counties, and on the west by Yavapai. Apache embraces a large area of the Colorado plateau, and its elevation above the sea level is from five to seven thousand feet, while some of its commanding peaks attain a height of over 11,000 feet. That portion of the county north of the Colorado Chiquito and the Rio Puerco, is composed of elevated table lands, isolated mountains, and deep and narrow canyons. In the northern end of the county is the remarkable plateau called the Mesa la Vaca, elevated about 1,000 feet above the surrounding formation. This is the great coal region of Arizona, which extends across the north-western portion of Apache county. This elevated region is covered by a growth of fine grass, crowned with stunted pines and cedars. Water is not plentiful. The extreme north-eastern corner of the county, through which flow the Rio de Chelly and its tributaries, is included in the Navajo Indian reservation. That part of Apache south of the thirty-fifth parallel is one of the best-timbered and watered portions of Arizona. The snowfall in this part of the territory is very heavy, giving rise to many beautiful, clear, mountain streams, which flow out through lovely valleys all the year round. The ranges of the Mogollon and the Sierra Blanco traverse this region, their summits covered with a heavy growth of timber, while the valleys and mesas are carpeted with rich and luxuriant grasses. The valley of the Colorado Chiquito contains fine farming land, and sufficient water for irrigation. Apache county has some of the best grazing lands in the Territory. In romantic and picturesque mountain scenery it is not equaled in Arizona.   •

GRAHAM.

Graham county, which · has just been organized from portions of Pima and Apache, is bounded on the north by Apache, on the east by New Mexico, on the south by Cachise, and on the west by Pinal and Gila. The Gila river flows through the center of the county, making a rich and fertile valley, which is being brought under a high state of cultivation. The Galiuro, the Pinaleno, and the Peloncillo ranges extend through the county south of the Gila, while north of that stream, the Gila mountains, the Sierra Natanes, and the Sierra de Petahaya cross its surface in every direction. The mountains are generally well wooded, while the broad valleys which lie between are covered with rich grasses, affording pasturage for large herds of cattle. The county is well supplied with water, and contains valuable mineral deposits near it eastern border.

GILA.

Gila county, called into existence by the last session of the Legislature, is bounded on the north by Yavapai, on the east by Graham and Apache, on the south by Pinal, and on the west by Pinal and Maricopa. It is a compact, mineral country, crossed in all directions by detached spurs and rolling uplands. The Pinal range is heavily timbered, and the whole county is covered with rich grasses. The Salt river flows nearly through the center of the county, while its southern border is washed by the Gila river. Gila is rich in gold, silver, and copper, and has, also, some fine cattle ranges. Its agricultural resources are confined to a narrow strip above the Salt river canyon, and the valley of the Gila and San Carlos, now included in the San Carlos Indian reservation.

PINAL.

Pinal county is bounded on the south by Pima, on the west by Maricopa, on the north by Maricopa and Gila, and on the east by Graham. South of the Gila, the county is made up of open, barren plains and isolated groups of rugged mountains. These plains are covered with rich gramma grasses, but devoid of water. The valley of the Gila, which flows through the county from east to west, is one of the most productive spots in the Territory, and yields large crops of grain and vegetables. The north-eastern part of the county is crossed by the Superstition, Mescal, and Salt River mountains. They are rich in mineral, though deficient in timber. The eastern corner of Pinal, south of the Gila, contains some fine farming and grazing land. The San Pedro flows through the county for nearly 40 miles, and its rich but narrow valley is under a high state of cultivation. Coal has also been discovered in this region, with every promise of permanency.

CACHISE.

Cachise county occupies the extreme south-eastern corner of the Territory. It was organized in 1881, from a portion of Pima county. It is bounded on the south by Sonora, on the west by Pima, on the north by Graham, and on the east by New Mexico. The massive chain of the Chiricahua runs through the county in the east, while the Huachuca, the Whetstone, the Dragoon, the Mule mountains, and the Galiuro ranges cross it from the north to south, in the west. All of these mountains are covered with pine, oak, and juniper, while the broad valleys that lie between, and the rolling table lands bear a generous growth of nutritious grasses. The San Pedro flows through the county from its southern to its northern boundary, carrying sufficient water to irrigate the rich and fertile valley that stretches along its banks. To the east of the Chiricahua range is the great valley of the San Simon, an immense extent of fine grazing land, with water to be found along its entire extent, within a few feet of the surface. The mountain ranges of Cachise are well watered, while the wonderful richness of their mineral deposits has attracted the attention of the entire country.

YUMA.

Yuma county, which comprises the south-western portion of the Territory, is bounded on the west by the Colorado river, on the north by Mohave, on the east by Maricopa and Pima, and on the south by Sonora. The Gila river flows through the county for nearly 100 miles, making in its course a fine valley, which is susceptible of high cultivation. The eastern portion of the county is composed of a high table land, with detached, rugged mountains crossing it in all directions. This table land is covered with coarse grasses, and affords excellent grazing, where water can be had. Many of the isolated ranges are known to be rich in minerals. That portion of Yuma county lying along the Colorado is traversed from north to south by parallel ranges of scorched and barren mountains, such as Castle Dome, the Plomosa, the Chocolate, and many other broken and detached ranges, nearly all of which are rich in the precious metals. Besides the farming land along the Gila, Yuma has a large and productive valley on the Colorado.

CHIEF TOWNS.

TUCSON.

Tucson, the county seat of Pima county, is situated on a sloping mesa on the right bank of the Santa Cruz river. It stands in a wide plain, surrounded on all sides by mountain ranges. It is about midway between the Gila river and the boundary line of Sonora, and is about 250 miles east of the Colorado river, and nearly 300 miles north of the harbor of

Guaymas, on the Gulf of California. It is situated near latitude 32° 20' north, and in longitude 110° 55' west from Greenwich. The early history of Tucson is involved in obscurity. It is generally believed that it was established as a Spanish military station to protect the mission of San Xavier del Bac, about the year 1694. Tucson remained a small and insignificant pueblo until the California gold fever of '49 and '50, when the rush of adventurers along the southern route to the golden shores of the Pacific infused new life into the sleepy old town. After the occupation of the country by the American troops, in 1855, Tucson became the most important point in the Territory, and its growth has been steady ever since. With the completion of the Southern Pacific railroad, the old pueblo has made rapid strides in population, wealth, and material prosperity, and contains, at the present time, between seven and eight thousand inhabitants, many of whom are Mexican. Tucson, in its general appearance, resembles a Spanish-American town. The houses, built of adobe, or sun-dried brick, are generally of one story, with flat roofs, and narrow doors and windows, with court-yards in the interior. The streets in the older part of the town are narrow and tortuous, and the houses make very little pretensions to architectural beauty. The advent of the railroad, however, has drawn hither an active, energetic American population, and the old order of things is being rapidly done away with. Tucson contains the largest mercantile houses in the Territory, who do a heavy trade with Sonora and the northern States of Mexico. The business of the town for 1880, amounted to over $7,000,000. The place contains some fine private residences, which would be a credit to any town on the coast. The Catholic cathedral is an imposing structure, built of brick and adobe. The Presbyterian church is a tasteful building of sun-dried brick. The Baptists have also a place of worship, and the Methodists have laid the foundation for a large and handsome edifice. Besides the public school, which is largely attended, the Sisters of St. Joseph have an academy for girls, with an attendance of nearly 100 pupils. A parochial school is also maintained with an enrollment of 285 pupils—160 males, and 125 females. The Odd Fellows, Masons, Knights of Pythias, Good Templars, and United Workmen, have flourishing lodges. Tucson has two banking-houses, four hotels, two breweries, two flouring mills, a foundry, and large mercantile establishments in every branch of trade. Three daily and weekly newspapers are published here. The Arizona Star, by L. C. Hughes, is a bright and able chronicle of the wants and resources of the southern country; the Arizona Journal, by F. B. Thompson, is a reliable and newsy exponent of public sentiment, and an active champion of the material interests of the country; the Arizona Citizen, the second oldest newspaper in the Territory, is conducted with ability by R. C. Brown, and is devoted to the vast and varied resources of Pima county and Southern Arizona. El Fronterizo, by Carlos Velasco, is published weekly, and supplies the Spanish-speaking population with the current news in their native tongue. The suburbs of Tucson afford some pleasant drives. San Xavier church is nine miles up the Santa Cruz, while Fort Lowell is at the base of the Santa Catarina mountains, seven miles away. The valley of the Santa Cruz, opposite Tucson, presents a beautiful appearance, with its green fields and groves of cottonwood. Situated on the main highway between the east and west, and on the direct route to the Gulf, with one railroad passing through it, and others projected, and with the rich mineral belt lying all around it, Tucson has every reason to feel secure in its future.

TOMBSTONE.

Tombstone, the county seat of Cachise county, is one of those mining towns which has sprung into existence, as if by magic, from the discovery of the wonderfully rich ore bodies which surround it on all sides. A little more than two years ago, the site of the present town was a desolate waste; to-day an active, energetic population of over 6,000 souls gives life and animation to its crowded streets. The town is built on a *mesa* at the southern end of the Dragoon mountains, nine miles east of the San Pedro river, about seventy miles south-east of Tucson and twenty-eight miles south of Benson, on the Southern Pacific railroad. It is situated near latitude 31° 30′ north, and in longitude 110° west of Greenwich. The first house was erected in April, 1879, and since then its growth has been remarkable. Surrounded on all sides by immense bodies of rich ore, Tombstone presents the appearance of a typical mining camp in the full tide of prosperity. The town is built of wood and adobes. It contains many fine business houses, a large and commodious theater and public hall, four large hotels, two banks, and numerous private residences, displaying both taste and comfort. It contains four churches: Methodist, a handsome edifice, Catholic, Presbyterian and Episcopal. It has one public school, which is largely attended, and also a private academy, which receives generous patronage.

Tombstone is the center of an immense area of rich mineral territory. It has a large and growing trade with the adjacent mining camps, and with Sonora. Its mercantile houses carry heavy stocks, and do a thriving business. Tombstone has two newspapers, the *Nugget* and the *Epitaph*, published daily and weekly. The former is the pioneer journal of the camp, and in its general make-up and the ability displayed in its columns, is worthy of the generous support it is receiving. It is conducted by H. M. Woods. The *Epitaph* is a live, newsy journal, devoted to the vast resources of the Tombstone region, and has worked incessantly to bring those resources to the attention of the outside world. Clum & Reppy are its proprietors. Water is brought to the town in iron pipes from the Dragoon mountains, sixteen miles away. A project is on foot to tap the cool springs in the Huachacas, twenty-one miles distant, which would supply the town with pure mountain water for all time to come. Tombstone is at present one of the most active towns on the Pacific coast. New buildings are going up constantly, while rich discoveries are being brought to light in the vast mineral belt which extends in all directions. Its future growth and prosperity is assured, and it promises yet to rival the metropolis of the Comstock in its most prosperous days.

PRESCOTT.

Prescott, the capital of Arizona, and the county seat of Yavapai county, is situated in a beautiful mountain glade, surrounded by the northern spurs of the Sierra Prieta. The town was laid out in May, 1864, and named "in honor of the eminent American writer and standard authority upon Aztec and Spanish American history." Its site is in latitude 34° 30′ north, and in longitude 112° 30′ west from Greenwich. The town has a beautiful situation, being surrounded by low hills, crowned with lofty pines, and covered with fine grasses. The streets are broad and laid out with the cardinal points of compass. In the center of the town is a large plaza, in which stands the county court-house, the finest structure in the Territory. It is built of brick and stone, two stories in height, with a mansard roof, crowned by a handsome tower. Prescott has the appearance of a homelike, Eastern town. Its buildings are of wood, brick, and stone. It contains the handsomest mercantile establishments in the Territory, many of which would be a credit to older and more pretentious communities. It is the center of an extensive mining, pastoral, and agricultural region, and has a large and prosperous trade. Besides its fine business establishments, Prescott can show many elegant private residences. It has a fine theater and a large public hall. Three saw mills are in constant operation near the town.

Prescott has one bank, a fine brick structure 72 by 29 feet, and two stories in height, two hotels, three breweries, fifteen mercantile establishments, and, like all frontier towns, numerous saloons. The town is situated about 5,500 feet above sea level, and possesses one of the most delightful climates on the continent; and with its pine-covered hills, green valleys, and beautiful gardens, is one of the most attractive towns on the Pacific coast. The Catholics, the Methodists, the Baptists, the Presbyterians and the Congregationalists, have handsome churches. A fine brick school-house, two stories in height, is one of the ornaments of the town. The Masons, the Odd Fellows, the Knights of Pythias, and the Foresters have flourishing societies. Two newspapers are published here, the *Arizona Miner*, the oldest newspaper in the Territory, and the *Arizona Democrat*. The former is conducted by C. W. Beach, and is untiring in its efforts to give publicity to the vast resources of Northern Arizona. The *Democrat* is owned and edited by Hon. Gideon J. Tucker, formerly of the Albany *Argus* and the New York *Daily News*. It is ably conducted, and justly appreciated for its devotion to the material interests of the Territory. The population of Prescott is about two thousand. With its charming situation, fine climate, and the varied resources which surround it, the town is destined to be a place of importance.

PHŒNIX.

Phœnix, the county seat of Maricopa county, is situated in the great Salt river valley, twenty-five miles above the junction of the Gila and the Salt rivers, and about two miles north of the latter stream, ninety miles south of Prescott, and twenty-eight miles north of the Southern Pacific railroad at Maricopa station. It is in latitude 33° 25' north and in longitude 112° west. The first settlement was made in December, 1870, in what was then a barren desert. By bringing the fertilizing waters of the Salt river over the plain, the valley has been made the most fertile and productive in the Territory. Phœnix is a beautiful town, with wide streets shaded with groves of cottonwood trees, and cooled by streams of water running through the principal thoroughfares. It is the center of trade for the productive farming region which surrounds it on all sides, and has a number of handsome mercantile establishments which do a prosperous business. It has three churches, Methodist, Presbyterian and Catholic, all handsome structures. The houses are generally built of adobe, as that material is found to be best adapted to this climate. A large, two-story brick school-house, is one of the chief adornments of the town. The Odd Fellows, Masons, Red Men, United Order of Workmen, and Good Templars have organizations here. The Maricopa Library Association is one of the most prosperous societies in the town. Two newspapers are published in Phœnix, the Phœnix *Herald* and the *Arizona Gazette*, the former by John J. Gosper, and the latter by McNeil & Co.; they are both well conducted, newsy journals, able exponents of the interests of the people and the resources of the Salt river valley, and are published daily and weekly. The population of Phœnix is about 1500, and is rapidly increasing. With its splendid water facilities and rich soil, with its fine farms, beautiful gardens, and shady groves, Phœnix is a handsome and a prosperous town, with a bright future before it.

GLOBE.

Globe, the chief town of Gila county and its county seat, is situated on Pinal creek, a tributary of the Salt river, about 120 miles north-west from Wilcox station on the Southern Pacific railroad, and about 90 miles north-east of Florence. It is a live mining town in the midst of a rich and extensive mineral belt. The place has a pleasant situation in the valley of Pinal creek, surrounded by rolling grassy hills, and backed by the lofty, pine-covered Pinal mountains to the south. The town is built principally of wood and brick, and presents a neat and attractive appearance. It has twelve mercantile houses, one bank, two hotels, a handsome Methodist church, a fine public school-house, two wagon shops, two drug stores, blacksmith shops, breweries, and several saloons. The town sprang up after the rich silver discoveries in this region in 1876. It has now a population of over 1,000, and a large and steadily grow-

ing trade with the mining camps adjacent. Globe has two
weekly newspapers, the *Silver Bell* and the *Chronicle*. The
former is conducted by Judge Hackney, and is a reliable and
consistent advocate of the wants and interests of Gila county,
and the Territory in general. The *Chronicle* is owned by W. H.
Glover, and is a staunch friend to its section and a credit to Arizona
journalism. Globe has an eligible situation in the center of a
vast mineral and grazing region, and is growing steadily.

FLORENCE.

Florence, the principal town of Pinal county, is situated
about 25 miles north-east of Casa Grande, on the Southern Pacific
railroad, 80 miles north of Tucson, and 45 miles south-east of
Phœnix. The town has a beautiful situation in the rich valley
of the Gila. It is surrounded by groves of cottonwood, clear
streams of water flow through every street, and beautiful gardens, where fruits and flowers grow luxuriantly, make it one of
the most attractive towns in the Territory. Its buildings are
principally of adobe, many of them tastefully adorned. Florence has several large business houses, two hotels, two commodious public schools, a Catholic church, a brewery, restaurants,
saloons, and two flouring-mills. The town was laid out in
1868, and has a population of 800, one third of whom are Mexican. It is the county seat of Pinal. The *Territorial Enterprise*, a weekly newspaper, is published here. It is an able and
industrious champion of the many resources of that portion of
the Territory. Florence is about 500 feet above sea level, in the
center of one of the finest bodies of agricultural land in the
Territory, and with rich mines north, south, and east, will
always be a prosperous town.

YUMA.

Yuma, the county seat of Yuma county, is situated near the
junction of the Gila with the Rio Colorado, and about twenty
miles north of the Sonora line. On a commanding bluff,
opposite the town, on the California side of the river, is Fort
Yuma, built on the site of a mission established here by the
Spanish fathers as early as 1771, and destroyed by the Yuma
Indians ten years later. The first settlement at the site of the
town of Yuma was made by Dr. Lincoln and others in 1849,
who established a ferry over the Colorado to accommodate the
thousands who flocked to the newly discovered gold region of
California. An outbreak among the Indians destroyed the
ferry and killed all the owners, except three persons. In 1850,
the ferry, was again started by Don Diego Jaeger and others.
This party were again attacked in 1851 by the Indians, who
compelled them to abandon their enterprise and retreat to
California. In 1852, Heintzelman and Stoneman (both of
whom afterwards rose to high commands in the civil war),
marched across the Colorado desert with a detachment of
United States troops, and established the post of Fort Yuma.
The ferry was again started, and the village of Arizona City

grew up around it. In 1864, Yuma was made the distributing point for the military posts in Arizona, and advanced rapidly in population and business. It contains several large stores, three hotels (one owned by the railroad company), a large wagon shop, blacksmith shops, saloons, etc. It has one public school with a daily attendance of 50. The Sisters of Charity have also a flourishing school at this place. The Territorial prison is situated here. It is a secure and roomy structure, built of stone, and situated on a bluff above the Colorado. The railroad company have built extensive shops at this point and give employment to a large number of men; they have also erected a fine bridge over the Colorado. The population is about 1,200. Yuma has two newspapers, the *Sentinel* and the *Arizona Free Press*. The former is conducted by J. W. Dorrington, and sets forth the local news of its section in an attractive manner. The *Free Press* is owned and edited by Samuel Purdy, Jr. It is an interesting journal, conducted with marked ability, and has done much to bring to notice the resources of Yuma county. Yuma's situation at the junction of the two largest streams in the Territory, the rich mining country which lies to the north and east of it, and its unrivaled climate for those troubled with lung diseases, will always insure its permanency and prosperity.

MINERAL PARK.

Mineral Park, the county seat of Mohave county, is situated on an elevated bench, on the western slope of the Cerbat range, 30 miles east of the Colorado river, and about 150 miles northwest of Prescott. The town is built mostly of adobe. It is the center of a rich mineral region. It was founded in 1871, and contains three stores, one hotel, one restaurant, one blacksmith shop, one public school, and four saloons. It does a thriving trade with the surrounding mining camps. The line of the thirty-fifth parallel railroad passes about ten miles east of the town. Present population about 300.

PINAL.

Pinal, a prosperous town in the county of the same name, is situated on Queen creek, about thirty-five miles north-east of Florence. The town is built of wood and a light-colored basaltic rock, which is found in abundance in the vicinity, and which gives the town a permanent and substantial appearance. The place has several large stores, two hotels, one bank (a handsome structure of stone), restaurants, saloons, blacksmith shops, and all the other branches of trade which are found in a prosperous mining town. Pinal has one church, and a public school which is well attended. The Pinal *Drill* is published here once a week by J. D. Reymert. It is a live journal, full of the local and general news of its section. The Odd Fellows have a fine hall and a flourishing organization in Pinal. The mill of the Silver King mining company is situated at this point, and many productive mines in the vicinity make Pinal a growing and prosperous town. Population about 600.

HARSHAW.

Harshaw is lively mining camp, situated in the northern spurs of the Patagonia mountains. It is built principally of wood. It has several mercantile establishments, who do a flourishing trade with Sonora and the adjacent mining camps. It has a population of about 600. The fine mill of the Hermosa mining company is located at this point. The place is about seventy miles south-west from Tucson. The town has a delightful situation, surrounded by the oak-covered hills of the Patagonia range. It is the center of a rich and extensive mineral region, and is destined to be a place of importance.

SILVER KING.

Among the other towns of note in the Territory, may be mentioned Silver King, which has been built up around the famous mine of the same name. It is situated about five miles from the town of Pinal, and is a thriving mining camp, having three stores, two hotels, and several saloons. Population about 250.

CHARLESTON.

Charleston, in Cachise county, is situated on the San Pedro river, about nine miles west of Tombstone. At this point are located the reduction works of the Tombstone Milling and Mining Company. The town has four stores, two hotels, besides blacksmith shops, saloons, etc. It is on the main road to Sonora, and does a large trade with that State. The population of the town is about 300.

GALEYVILLE.

Galeyville is a lively mining town, situated on Turkey creek, on the eastern slope of the Chiricahua mountains. It is twenty miles south of the Southern Pacific railroad, and thirteen miles west of the New Mexican line. It has a beautiful situation, surrounded by groves of oak. The town was laid out in November, 1880, and has a population of about 400. There are six stores, four restaurants, two blacksmith shops, two feed and livery stables, three butcher shops, thirteen saloons, barber, boot and shoe shop, etc. The town is surrounded by a rich mineral belt, and promises to become a place of importance. The country in the vicinity has an abundance of wood, water, and fine grasses.

ST. JOHNS.

St. Johns, the county seat of Apache county, is situated on the Little Colorado river, about two hundred miles in a direct line east of Prescott, and about twenty miles west of the boundary line of New Mexico. It is in the center of a rich agricultural and grazing region, contains a population of 700 souls, a large portion being Mexicans. The town is on the direct road from Fort Wingate to Fort Apache, and about forty miles south of the Atlantic and Pacific railroad. A large and commodious court-house has recently been erected. The town does a large trade in grain and wool, and has four stores, saloons, blacksmith shops, etc.

SAFFORD.

Safford, the county seat of Graham county, is on the Gila river, near Camp Thomas, and in the center of that rich farming region known as the Pueblo Viejo. The town is steadily growing, has a population of about 300, and has a large trade with the agricultural region which surrounds it. It contains several stores, a hotel, saloons, etc. With its unrivaled farming and grazing resources, Safford is destined to become a large town.

MINING RESOURCES.

The very name of " Arizona " is suggestive of streams yellow with golden sands, and mountains glittering with virgin silver. Popular belief has long considered this region as a synonym for marvelous mineral wealth, and long before that wealth was proved to have an existence, tradition and story had woven about the name a glamour of golden fancies, which modern enterprise and modern energy are at last about to turn into solid facts. The first mention of the Territory in history is connected with the search for the treasures supposed to be collected in the Cities of the Bull; but although the expedition did not result so successfully as a similar one in an earlier age, which sought and found the Golden Fleece, it was indirectly the means of leading to the discovery of the buried treasures which underlie the mountains and valleys of this wonderful land. The hardy adventurers who followed Coronado little dreamed that the mountains, plains, and mesas, which they passed over in their wearisome journey to " Cibola," contained riches, which would make the fabulous wealth of the Moqui cities appear mean and insignificant. It has remained, however, for a later age and another race to bring to light this vast wealth, and send it forth to benefit mankind, and enlarge and enrich the trade and commerce of the globe. The Territory of Arizona is one vast mineral field; from the line of Utah on the north, to the Mexican border on the south, and from the Colorado of the west, to the boundary of New Mexico, mineral is found in nearly every mountain range, and in every isolated peak. Nowhere on the continent is there such an extensive distribution of the precious metals. While in other mineral-bearing States and Territories the deposits are confined to certain well-defined limits, in Arizona no such distinction prevails. It would appear as if nature had here, in a prodigal mood, scattered her treasures with a lavish hand, and neglected no portion of her chosen mineral domain.

In the richness and variety of its ores, Arizona is also distinguished from the mining regions of the west. This predominating feature of the country was noted at an early period in its

3

history. No mining State or Territory has yielded such masses of pure silver, and few have equaled the wonderful gold deposits of Antelope Hill. To Arizona belongs the honor of producing the largest nugget of native silver ever found— 2,700 pounds. This mass of pure metal was confiscated by Philip V., and taken to Madrid. The mine was also declared government property, but it does not appear that the royal robber ever derived much benefit from it. The many rare and beautiful combinations in which silver is found make Arizona the favorite field of the mineralogist, while the ease and simplicity by which these ores are reduced commends itself to investors and to metallurgists alike. Pure native silver, chlorides, ruby silver, bromides, silver glance, sulphides, carbonates, and sulphurets are the most generally distributed of the silver ores, but there are many other varieties peculiar to the Territory, which space will not permit to mention here. Gold is most generally found in its matrix of quartz. It sometimes occurs in conjunction with pyrites of iron, copper, and lead, and is met with in its pure state in creeks and gulches in all portions of the Territory. Copper is found in red and black oxides, as a green and blue carbonate, sometimes as a sulphate, and often in its native state. Silver ores in Arizona, which assay into the thousands, are of common occurrence, and create no comment. Large quantities of ore going from $5,000 to $10,000 per ton, have been shipped from the Territory, and several mines are steadily producing "rock" that will go from $15,000 to $20,000 per ton. These are simple facts which can not be gainsayed.

Probably no portion of the mining domain possesses so many natural advantages for the successful working of ores. Wood and water are abundant in nearly all of the mineral-bearing mountain ranges, and in places where water is scarce at the surface, a sufficient quantity is found by sinking a short distance. The climate of the country can not be excelled. Work can be prosecuted all the year round. While mountains of snow and intense cold retard operations in other States and Territories, Arizona's equable climate is specially adapted to out-door operations, even in the middle of winter. This fact alone is worthy the careful consideration of men desiring mining investments. The old shafts and tunnels which have been discovered in various parts of the Territory, show that the Spanish explorers and the early missionaries had proven the richness of Arizona mines, and had, in their crude way, worked them successfully. The almost indisputable evidence which an earlier race of miners have left in several of the gold-bearing streams of the Territory, proves conclusively that the people who once occupied this land, and whose origin is lost in the mists of conjecture, delved for the precious metals in this region—at once the oldest and the newest portion of the American Union. The same difficulties which obstructed the operations of Toltec and Spaniard has also stood in the path of their Anglo-Saxon successors. Isolation and savagery have retarded

Arizona's development. These two words express the causes which have prevented the country's advancement, and deprived her of the position which she is soon destined to attain—the leading bullion-producer on the globe.

But now that the savage has succumbed to his destiny, and the mountains and valleys which once resounded with his war-whoop, re-echo the music of civilized industry; now that the Demon of Isolation, whose shadow hung like a funeral pall over the land, has been driven to more distant fields by the shriek of the locomotive, Arizona is rapidly coming to the front as the most promising mineral region in all North America. An army of prospectors are swarming through her valleys and mountains; new discoveries are constantly being made; mills and furnaces are going up; the yield of bullion is steadily on the increase; capital is seeking investment; railroads are penetrating in every direction, and henceforth the career of Arizona is to be onward and upward. The scope of this work will not admit of a detailed or elaborate description of every mining district in the Territory. It is believed, however, that in the following brief summary of the leading camps, enough will have been shown to prove all that we have claimed for the richness and extent of the mineral field; the natural appliances for the reduction of ores, and the unrivaled opportunities which the country presents for the investment of capital.

CACHISE COUNTY.

In the fall of 1877, Mr. A. E. Sheiffelin, an active and industrious prospector, was stopping at Camp Huachuca. He made frequent trips into the hills now embraced within the limits of Tombstone, searching assiduously for "float" and "croppings." Bands of renegade Indians roamed in the country east of the San Pedro at that time, and the whole region, which had once been the chosen ground of the famous Cachise and his band, was marked with the graves of white victims, who had been murdered within its "dark and bloody ground." Sheiffelin was admonished that he would find a "tombstone," instead of a "bonanza," beyond the San Pedro, and would add another to the many who found bloody graves among its lonely hills. The indomitable prospector paid no heed to these warnings, and his pluck and energy met with their just reward. In February, 1878, he discovered the Lucky Cuss, Tough Nut, and other mines which have since attained a national reputation. In remembrance of the solemn joke, he named the district "Tombstone." The great richness and extent of the new discoveries soon spread far and wide, and thousands rushed to the Silverado of the south-west. An army of prospectors swarmed over the hills, many other valuable discoveries were made, a city sprung up as if by magic, mills and hoisting-works were erected, bullion began to find its way out of the camp, and to-day, a little more than three years after its discovery, Tombstone can show a population of 7,000 souls, and is one of the most prosperous mining camps in the western country.

As near as can be ascertained, the mineral belt of Tombstone extends nearly eight miles east and west, and about five miles north and south. On the western edge of the district, along the San Pedro river, silver had been discovered as far back as 1859, but the hostility of the Indians prevented any development. The country in which the mines of Tombstone are situated may be described as a series of rolling hills, which have a gradual ascent until they merge into the Mule mountains' on the south, and stretch away in an undulating plain to the Dragoon range on the north. The geological formation of the district presents many features worthy of study. Porphyry appears to be the predominating rock, though a capping of lime overlies the leading mines of the camp. Quartzite is found everywhere, and a granitic formation is met with on the western edge of the district. As depth is attained, the surface lime disappears and porphyry and quartzite constitute the country rock. A notable feature of the Tombstone mines is the size of the veins and the ease with which the ore is reduced. The silver occurs as a chloride with very little base combinations, and can be worked by pan process, to 90 per cent. and upwards. The cost of extraction is merely nominal, and the facilities for reduction are all that could be desired. The present output of bullion is over $500,000 per month, from 140 stamps. This yield is being steadily increased, and valuable paying properties are being added to the list of bullion producers every month. It is estimated that the bullion yield for the present year will amount to $7,000,000. This is a good showing for a camp a little over three years old, which did not drop a stamp until June, 1879. The daily output of ore at the present time is about 500 tons. Fourteen of the leading mines have complete hoisting-works with the latest improved machinery. Water has been struck in several claims at a depth of between 500 and 600 feet, but the inflow is as yet very light, and no difficulty is experienced in getting rid of it. There are over 3,000 locations in Tombstone district. In this brief sketch there are doubtless many promising properties deserving of notice besides those mentioned, but space will not admit of a separate description of each.

The Tombstone Gold and Silver Mining Company own the Lucky Cuss, the East Side, Tribute, and Owl's Nest. This group constitutes one of the most valuable properties in the district. The Tough Nut, the leading mine, is thoroughly opened by shafts, drifts, winzes, and open cuts. Immense ore bodies, sometimes 20 feet in width, are met with. The ore is found in spar and quartz, and is said to average $100 per ton. The company have two mills on the San Pedro, one of 10 and another of 20 stamps. It has paid dividends from the start, and has a large surplus on hand. This is the first organized company in the district. It employs about 125 men, and its production of bullion up to date, is said to be about $1,000,000. The Grand Central Company's property is embraced in a claim 1500 feet in length and 600 feet in width. It is incorporated under the laws of Ohio, with a capital of $10,000,000, divided into 100,000 shares. It is a magnificent property. The vein is from 8 to 12 feet wide, and runs from $80 to $100 per ton. The main shaft is down 500 feet, with three levels—500, 1100, and 600 feet, respectively. The reduction works consist of 30 stamps on the San Pedro, which are kept constantly at work. While only in operation a few months, it is estimated that $500,000 has already been produced. Regular dividends are declared, and the property is steadily increasing in value as depth is reached. The Western Company own the Contention, one of the first locations in the district, which has produced a large amount of bullion. The property joins the Grand Central on the north. The writer was not permitted to see the mine, and therefore can say nothing definite about its present condition.

The Girard has a shaft 400 feet in depth and a vein from 4 to 6 feet in width. The ore is of high grade and has milled $100 per ton. The property is incorporated in Jersey City with a capital of $2,000,000, divided into 200,000 shares. The company have put up fine hoisting-works and will soon have a mill in operation on the San Pedro. The Head Center embraces 1,300 feet in length and 500 feet in width. It is incorporated under the laws of the State of California with a capital of $10,-000,000, in 200,000 shares. The vein averages from 4 to 8 feet, yielding about $70 per ton, about 45 per cent. of the bullion being gold. The company own a 10-stamp mill near Contention City. The main shaft is down 600 feet. The first level is 500 feet, the second 400, and the third 500. Hoisting machinery of the most improved pattern has been erected. The Vizina is incorporated under the laws of the State of New York on a basis of $5,000,000 and 50,000 shares. The mine is opened by three shafts, the deepest being about 400 feet. It is the intention of the, company to erect a mill at an early day. Meanwhile the mine is being thoroughly opened. Over $200,-000 has already been taken out from ore worked in a custom mill. Fine hoisting machinery has been erected, and the work of development is pushed forward steadily. The Empire is bounded on the south by the Sulphuret and the Girard. It is incorporated under the laws of Massachusetts. The main shaft is down 450 feet and has struck a large body of high-grade ore. A hoisting engine, with a capacity to sink 1,200 feet, has been put up, and this valuable property is being thoroughly opened. The Sulphuret adjoins the Empire and the Head Center. It is incorporated under the laws of Pennsylvania. Its main shaft is down 600 feet. It has a fine location; has first-class hoisting-works, and is being opened in a systematic manner. The Bob Ingersoll, one of the most valuable claims in the district, shows 5 feet of ore that will mill $100 per ton. It has a shaft down 200 feet, and is steadily improving as it is being sunk upon. This mine is incorporated. The Sydney is a fine-looking property with a vein 12 feet wide, 4 feet of which is ore that goes from $50 to $100 per ton. The mine is owned by

San Francisco parties. The Grand Central South has a shaft 250 feet in depth. It is a large vein adjoining the Grand Central, and is considered by many the coming mine of the camp. It is incorporated in San Francisco.

The Tranquillity joins the Empire and the Girard on the west. It has expensive hoisting-works, and is showing some very fine ore. None of the stock of this mine is on the market. The Flora Morrison is bounded on the east by the Grand Central. It is incorporated under the laws of Pennsylvania; 250,000 shares, $2 per share. It has a shaft 300 feet deep, besides drifts, cross-cuts, and winzes, and is showing fine ore. The Way Up has a shaft 300 feet, and is producing ore of a high grade. It is incorporated in New York; 150,000 shares, $10 per share. The Lucky Cuss, one of the first locations in the district, has a shaft 300 feet, and over 500 feet of drifts and cross-cuts. It has produced some of the richest ore ever taken out in the camp, and yielded about $50,000. The Sunset, south of the Lucky Cuss, has produced over $50,000. The Wedge shows a vein 3 feet wide, of high-grade ore. It has a shaft 100 feet deep, which is steadily pushed downwards. The mine is incorporated. The Gilded Age adjoins the Goodenough, and embraces a large portion of the town site. It has one shaft down 100 feet, which has produced rich ore. The Mountain Maid has a vein from 2 to 4 feet, and runs from $50 to $300 per ton. It has 3 shafts, the deepest being 200 feet. Like the Gilded Age, it extends across the town site. Among the many other claims in the immediate vicinity of the town, may be mentioned the Cincinnati, Grand Dipper, Naumkeg, Hawkeye, Plum, Rattlesnake, Wide West, Topaz, Omega, Omaha, Alpha, Prompter, Sunrise, Parallel, Little Wonder, Revenue, Survey, Defense, and hundreds of others worthy of mention here if the space permitted. Many of these claims are steadily and surely developing into fine paying properties.

In the western portion of the district are several well-defined and valuable mines showing rich ore, and large veins. The following are the most prominent: Owl's Nest, carrying 3 feet of ore that goes from $50 to $80 per ton. This claim has 3 shafts, the deepest being 100 feet. It is owned by the Tombstone Mining Company. The Junietta has a 2-foot vein assaying $150 per ton. The deepest shaft is 100 feet. The Silver Bell has a shaft 50 feet, and carries ore worth $100 per ton. The Stonewall has a large ore body that has yielded $75 per ton. It has a shaft 120 feet. The Monitor is a 6-foot vein of free-milling ore, going $40 per ton, with a shaft 120 feet, in a granite formation. The Merrimac has 4 feet of ore that has milled $60. It has two shafts 60 feet each, and one 40 feet. Both these claims belong to the Monitor Mining Company, an Eastern incorporation. The True Blue is a 2-foot vein of $100 ore, with a shaft 200 feet. The Lucknow has a shaft 50 feet, and has ore that averages $50 per ton. The Delhi, Miami, Franklin, Randolph, Red Top, Argenta, Three Brothers, and many others, are in this neighborhood, and are well worthy of notice.

Three miles from the San Pedro, is another group of mines which are producing remarkably rich ore. The Bradshaw, in its bullion yield and development, is the best known of these claims. It is a large vein, carrying ore that works from $80 to $100 per ton. It has been sunk to a depth of 400 feet; has improved hoisting machinery, and has already produced nearly $50,000. It is owned by an incorporated company in San Francisco. A 10-stamp mill is nearly completed, and the mine promises to be one of the regular bullion-producers of the district. The Alkey is a 4-foot vein, producing ore worth $100 per ton. It has a 50-foot shaft. The Bronkow, the first location in the district, is a vein 6 feet wide. It has a shaft 60 feet deep. Continual litigation has retarded the development of this property. In this necessarily brief *résumé*, full justice can not be done to the immense silver veins of Tombstone district. The salient points only have been given; but to have a proper conception of the size, richness, and extent of the veins in this wonderful camp, a personal examination is necessary. It is safe to say that nowhere on the coast have there been found ore bodies larger, richer, or more extensive. There are hundreds of fine prospects as yet undeveloped, which give every indication of being valuable, and which offer admirable opportunities for investment.

CALIFORNIA DISTRICT is situated in the Chiricahua mountains, twenty miles south of the Southern Pacific railroad, near the New Mexican line. The country is well wooded, and water is abundant. A thriving camp has sprung up, and many rich and valuable mines have been discovered. The ores are generally smelting, carrying much horn silver. The veins are large and well defined. Its proximity to the railroad and its abundance of ore, make Galeyville one of the most promising camps in Cochise county. The following are among the leading mines of the district: The Texas, the principal mine of the camp and the first discovered, is a large vein from 4 to 30 feet wide. The ore is a galena and chlorides, and averages about $40 per ton. A shaft 300 feet, and 3 tunnels, 250, 30, and 40 feet, respectively, expose large ore bodies. A 30-ton smelter has been erected and is now fairly under way. The Texas Milling and Mining Company are the owners of the property, which includes ten other mines in the same group. The Continental shows 2 feet of ore, assaying $100 per ton, principally chlorides and bromides. It has a shaft 60 feet and a cut 30 feet. The Cashier shows 4 to 6 feet of ore, and assays from $30 to $200 per ton. There are many other claims in this district looking well and producing good ore, which must be omitted here, but which are well worthy inspection by those who are desirous of investing in desirable mining properties.

TURQUOISE DISTRICT.—This district is situated about 18 miles north-east from Tombstone, at the southern end of the Dragoon mountains. There is plenty of water, and sufficient wood to last for years. The ores are smelting, easily reduced, and running from $40 to $300 per ton, with an average of about

$80. The Mono mine shows a vein from 2 to 6 feet wide. It is a carbonate ore, which will smelt readily. Assays go about $80 per ton, on an average. The mine is opened by about 500 feet of shafts and drifts, and shows fine ore in every opening. It is owned by a New York company. The Defiance and the Dragoon claims are also owned by New York parties. The former shows from 2 to 20 feet of carbonate ore, which will average about $80 per ton. There are several hundred tons on the dump. Reduction works will soon be erected on this property. The Dragoon has a 60-foot shaft showing a 4-foot vein that goes about $80. The Bell is the south extension of the Defiance. It is a 4-foot vein, looking well. The Challenge and the Tom Scott are also very promising veins, with ore that goes $75 per ton. The Star and Bodie claims are two of the best properties in the district. The Star has a shaft about 60 feet deep, all the way in ore that runs about $60 in silver and $12 in gold per ton. The Bodie has a 70-foot shaft, with a 2-foot ledge that averages $80 per ton in silver. With its favorable surroundings and fine ore bodies, Turquoise is destined to become a prosperous camp.

Dos Cabezas or "Two Heads" district is situated in the Chiricahua range, in the north-eastern portion of Cochise county. Its ores are gold-bearing, carrying some silver, and its ledges are large. It is favorably situated near the line of the Southern Pacific railroad, and has plenty of wood and water. The following are the principal mines in the district: Silver Cave has three veins, 7, 5, and 3 feet wide, respectively. The yield per ton has been $35. Several shafts, drifts, and other openings have been made on this claim, and nearly $5,000 has been taken from it, the ore being worked in arrastras run by steam. The Juniper is a 6-foot vein, carrying gold and silver. The ore assays $150 per ton. About $6,000 has been taken from this mine, the ore being worked in arrastras. The Silver Cave South, has 4 feet of ore that assays $50 per ton, and has several openings. The Galena Chief shows 3 feet of ore, assaying $50 per ton. The Murphy is a 4-foot vein, averaging $50 per ton. The Bear Cave has nearly 4 feet of ore that goes $80 per ton. The Greenhorn is also a 4-foot ledge, running $50 per ton. There are many other promising prospects in this camp well worthy of mention. With the erection of a 10-stamp mill, which is already on the road, Dos Cabezas will give a good account of itself.

Swishelm District.—This district is situated in the Pedrogosa mountains, in the south-east corner of Cochise county. Its ores are a carbonate. The veins are large, and the facilities at hand for smelting, good. A St. Louis company is now operating in the district with satisfactory results.

Hartford District.—This district is situated in the southern end of the Huachuca mountains. It has abundance of fine water, and some of the best pine timber in the Territory. Most of the lumber for Tombstone comes from this point. The ores are a copper and a carbonate of silver, assaying from $15 to $60 in copper, and from $20 to $80 in silver. Some very fine properties have been opened up. The Undine, Mountain View, Lone Star, and IXL, are the principal mines. Several sales have been made, and with the unsurpassed advantages of wood, water, and magnificent climate, Hartford district is certain to become an important mining center. There are several other points in the Huachuca range that show fine prospects, and also in the Whetstone mountains, west of Tombstone.

Copper.—Besides its veins of silver and gold, Cachise county has also some of the largest and most valuable copper mines to be found in the Territory. At Bisbee, some twenty miles south of Tombstone, are found some of the richest copper mines in the United States. The veins are large, the grade high, and the appliances at hand for reduction can not be excelled. The mines are about sixty miles from the railroad at Benson, and about twenty miles from the Sonora line. The Copper Queen, the leading mine of the camp, is an immense mountain of ore. It has been explored 160 feet in length by 150 in depth, and 120 feet in width, and as far as the explorations have extended, rich ore has been encountered everywhere. The claim is 1,500 feet long, and 600 feet wide. Two 30-ton smelters are kept running steadily, and the daily output is about 13 tons of pure copper. The ore is a carbonate and a black and red oxide, and averages about 22 per cent. The claim has been opened by 700 feet of shafts, drifts, and cross-cuts, and has already yielded over $600,000 worth of copper. The property is owned by an incorporated company, with headquarters in New York. The Neptune company own nine claims, the most prominent of which is the Neptune, which shows ore going 24 per cent. This company are making preparations to erect a smelter on the San Pedro river, fifteen miles distant. The Twilight shows a 6-foot vein of red oxides, carrying 25 per cent. pure copper, and is opened by a 70-foot shaft. The Holbrook has a 10-foot vein of red oxides, but has little work done on it. The Copperopolis shows a 5-foot vein and a 40-foot shaft. The Atlanta carries 25 per cent. ore, and is opened by a 45-foot shaft. The Copper King is the western extension of the Copper Queen. It is a large vein, showing good ore. The Golden Gate, Ohio, Copper Prince, Cave, New York, Galena, Garfield, Bounty, Black Jack, and Dreadnaught are all fine prospects, although but little work has been done upon any of them. Bisbee, besides its immense copper veins, has silver and gold also. It is one of the most eligibly situated camps in Southern Arizona, and has a bright future before it.

PIMA COUNTY.

This county is the oldest mining region in the United States. At what time the first discoveries were made by Europeans is not clear, although it is believed that the Jesuit missionaries operated here as early as the latter part of the seventeenth century. By the middle of the eighteenth century mining was prosecuted vigorously in the Baboquivari, the Santa Rita, Arivaca,

Oro Blanco, Patagonia, and at several other points in the
county. From the many old shafts and tunnels which have
been discovered, it is evident that the industry was carried on
extensively. In this region was found the famous "Planchas
de Plata," or "planks of silver," which yielded nearly five
tons of the pure metal. Many of the rich mines which were
worked in those days, have not been found, although the most
diligent search has been made. The abandonment of the missions in 1828, and the hostility of the Apaches, almost put a
stop to mining in Arizona, and it was not until some time after
the country came into the possession of the United States, that
it was resumed. Several companies were then organized, and
a great deal of bullion taken out. The constant raids of the
savages, and the withdrawal of the troops, on the breaking
out of the civil war, almost put a stop to all work, and not
until the Indians were subdued, in 1874, did the mining industry of Pima county take a fresh start. This industry has
received a wonderful impetus by the building of the Southern
Pacific railroad. Millions of dollars have been invested; new
districts have been organized; an army of prospectors has in-
vaded the country, and many valuable discoveries have been made.
Gold, silver, copper, and lead, are found in every mountain
range in the county. With the exception of the region known
as the Papagueria, wood and water is abundant everywhere.
The richness of its ores and the size and permanent character of
its veins, have given Pima county a reputation second to no
portion of Arizona.

HARSHAW DISTRICT.—This district is about 70 miles south-east
of Tucson, in the Patagonia mountains. The hills are covered
with oak and juniper, while the water supply is sufficient for
the working of ores. The camp is about 50 miles south of the
Southern Pacific railroad. The Hermosa is a large lode of free-
milling ore. The vein is from 8 to 12 feet wide. The ore is a
chloride and horn silver. One of the most complete 20-stamp
mills on the coast is kept steadily at work, crushing about 80
tons per day. The yield of bullion up to date has been over
$700,000. The mine is opened by a tunnel 700 feet in length,
cutting the vein 300 feet below the croppings. A shaft has
been sunk 100 feet below the level of this tunnel, and the mine
is thoroughly opened by drifts and cross-cuts. The Hermosa
is one of the leading mines of the Territory. The Hardshell is
a short distance west of the Hermosa. It shows a vein from
10 to 12 feet wide, of the same character of ore as the latter-
mine. It is opened by a shaft 50 feet deep, and by several
cross-cuts. The Hardshell gives every promise of becoming
one of the first mines of Pima county. The Trench is one of
the old mines worked by the early missionaries, and carries
some ore of a high grade. It shows a vein from 3 to 4 feet
wide, carrying sulphurets of silver. The main shaft is down
400 feet, and several levels have been opened. Steam hoisting-
works of the latest pattern have been erected. The Alta, south
of the Hermosa, is opened by several shafts, and shows a large

ore body. It is owned by Eastern parties. The Blue Nose, the American, the Independent, and many other fine prospects in this camp, show good ore and large veins.

WASHINGTON CAMP is about nine miles south of Harshaw, and was formerly known as the Patagonia district. It is in the southern end of the Patagonia mountains, and has a delightful situation, being in the midst of a heavily timbered region. The Santa Cruz river, four miles distant, affords an inexhaustible • supply of water. The district contains large veins of low-grade ore, carrying a heavy percentage of lead. The Davis is an immense vein of carbonates, being in places 20 feet wide. It has been opened by several shafts, the deepest being 160 feet, and also by drifts, cross-cuts, and winzes. The vein throughout all its workings shows large quantities of ore. The property is owned by the Patagonia Mining Company, who have erected a furnace on the Santa Cruz. The Belmont is one of the oldest locations in the district. It is three miles from the Sonora line. It has a shaft over 100 feet, and a cross-cut at the bottom showing 30 feet of carbonate ore, carrying considerable iron. The San Antonio is also an old location. It is opened by three shafts, the deepest being 60 feet. It shows a large body of ore similar in character to the Belmont. The Holland is another large body of smelting ore. A shaft has been sunk nearly 100 feet, showing a strong vein in the bottom. The Washington is a vein, in places 30 feet wide. The ore carries iron and copper pyrites, and requires to be roasted. It is opened by several shafts and drifts. What is known as " Washing Pool mines " embrace the Grasshopper, St. Louis, Chicago, Ella, Ohio, Columbus, Blue Jay, and many others. They are all large veins, carrying ore of a good grade, though mixed with much base metal. The "Old Mowry mine " is four miles north of Washington camp. Before the breaking out of the civil war, the mine was worked by Lieutenant Mowry, giving employment to 400 men (principally Mexican). Large smelting works were erected, but the tall brick chimney is all that remains of the ruin. During the war the Apaches destroyed the building and machinery. The old shaft is down 350 feet. The ore is easily smelted, and carries from 40 to 60 per cent. lead. The mine is now owned by parties in Tucson. The Redoubtable, Pensacola, Pelican, Chico, Thurman, and scores of others show large veins, and many of them have shafts from 60 to 70 feet. Washington Camp is favored beyond most districts in its natural facilities for ore reduction. This, together with its immense veins, should yet make it one of the leading districts in the Territory.

TYNDALL DISTRICT is situated south of the high peaks of the Santa Rita mountains and about sixty-five miles from Tucson. The ores of the district are generally of a good grade, but the rich ore bodies are not large. This camp has suffered from bad management by unscrupulous speculators. The mines are favorably situated near the Santa Cruz, while plenty of wood is found on the mountain sides. The Josephine is a vein 5 feet wide, of free-milling ore that has worked $60 per ton. It is opened by a shaft 75 feet deep. The Emma shows a vein 6 feet wide, some of which has yielded $100 per ton. It has a 50-foot shaft. The Magnolia has 3 feet of milling ore that assays $70 per ton. It has a shaft 30 feet. The North Star has a shaft 50 feet, and shows a 5-foot vein that has yielded $50 per ton. The Bonanza is opened by an 80-foot shaft and shows over 7 feet of fine smelting ore. The Dayton is a 4-foot vein, some of which assays $150 per ton. There is a shaft on this mine 100 feet in depth. The Bushnell shows a 6-foot vein and a shaft over 100 feet. Some ore from this claim assays $300. It is a smelting ore. The Lost mine has a shaft 150 feet and a 4-foot vein assaying $60 per ton. The Major has a tunnel 100 feet in length. Its vein is 5 feet wide, and it has produced rock that has assayed $700. The Jefferson is a large body of smelting ore, nearly 7 feet wide. It has a shaft 150 feet in depth. The Red Cloud is a 3-foot vein of free-milling ore. It is opened by an 80-foot shaft, and has produced some very rich rock. The Laura, Happy Thought, Gold Tree, Helvetia, Red Oak, Hidalgo, Cachise, Hamilton, Alcalde, Davis, Crown Point, and many other promising properties, are found in Tyndall district.

THE AZTEC DISTRICT is really a continuation of the Tyndall. The character of the ore is the same and the formation similar. The veins are large and well defined, and can be traced for a long distance. The same causes which have retarded the development of the Tyndall district have also operated here. Among the claims which deserve mention, are the Empress of India, San Ignacio, Old Salaro mine, Rosario, Las Cruces, Ricard, Anahuac, Toltec, Coronado, Henry Clay, Apache, Santa Rita, Hidalgo, Seneca, La Salle, Juarez, and many others.

ARIVACA DISTRICT.—This district is about 65 miles south of Tucson. Mining was carried on in this region long before the settlement of the country by the Americans. The camp has a delightful situation, a fine climate, and is possessed of abundance of wood and water. The formation is granite and porphyry. The Con. Arizona is owned by the Consolidated Arizona Gold and Silver Mining Company. It is opened by a main shaft 200 feet in depth, and by levels and drifts. The ore is a chloride, which mills freely. The vein is from 3 to 5 feet wide, and the yield has been about $100 per ton. A complete 10-stamp mill has been erected on the property, and also steam hoisting-works. The vein has fine walls, and gives every indication of being a permanent fissure. The company own three other claims on the same vein, among which the Silver Eagle has the most development. It has a shaft 78 feet, and shows a 4-foot vein that assays $75 per ton. The Albatross is a large body of sulphuret ore that gives an average assay of $80 per ton. It is a new discovery, and has been opened by a shaft 60 feet in depth. The Arkansas is a 4-foot vein, carrying chlorides and sulphurets of silver. The ore assays $100 per ton. The mine is opened by a 150-foot shaft and a tunnel 300 feet. The Dos Amigos shows a vein 3 feet in width that gives $80 as an average assay. It has a shaft 100 feet deep. The Idaho is a large vein, carrying ore that goes $30 per ton. It is opened by an 80-foot shaft. The Union shows a 4-foot vein of free-milling ore assaying $50 per ton. A shaft 120 feet deep has been sunk on the claim. The Fairview is a 4-foot vein carrying ore that goes $40 per ton. It has a shaft 130 feet deep. The Relief has a shaft 55 feet, and a vein 4 feet wide, going $50 per ton. The Postboy shows a vein 2 feet in width, of carbonate ore, that has assayed $100 per ton. It has a shaft 30 feet deep. The Longarine is opened by two shafts 100 and 80 feet, respectively, and by 300 feet of drifts and winzes. The ore is free-milling, assaying $80 per ton. The Clipper, Tennessee, Alpha, Grand Republic, Arion, Black Eagle, Blue-jay, Mentor, and Arivaca are among the many promising prospects of this district. No portion of Pima county presents a more inviting field for investment. The famous Cerro Colorado mine is about ten miles north of Arivaca. It was worked extensively, under every disadvantage, before the breaking out of the civil war, and has produced, it is said, nearly $2,000,000. The constant attacks by the Apaches compelled the abandonment of the property, and the buildings and hoisting-works were destroyed by the savages. It is now owned by the Arivaca Milling and Mining Company. The vein is not large, but the ore is of a high grade. Bounding Arivaca on the west is the Baboquivari range, which has been mined by the old missionaries in the early days, and contains some large veins of rich ore. The Oro Fino is a vein nearly 8 feet in width, assays from which give $50 per ton. It has a shaft 60 feet. The Black Hawk is an 8-foot vein of base metal, that gives $50 per ton. A 60-foot shaft has been sunk on the property. The Silver Chief shows 4 feet of ore that assays $60 per ton. It is opened by a shaft 150 feet deep.

Oro Blanco.—This camp is seven miles south-east of Arivaca. The country rock is generally porphyry. The ores are mostly carbonates and free-milling. Wood is plentiful. The ores carry gold and silver. The Warsaw is a vein from 3 to 4 feet wide. Ore from this mine has worked $80 per ton. It is opened by a 300-foot shaft, besides drifts and cross-cuts. A ten-stamp mill and roaster have been erected on the property. It has produced over $25,000. The Alaska is a 4-foot vein, carbonate ore, assaying $70 per ton. It has a shaft 150 feet deep, and a 200-foot tunnel. The Peelstick has a shaft 170 feet deep, has a 4-foot vein, and assays $60 per ton. The Yellow Jacket has a shaft 120 feet deep, and 400 feet of drifts and tunnels. It shows a ledge 3 feet wide — gold quartz. A ten-stamp mill has been erected on the mine, and a considerable amount of bullion taken out. The Montana is a large ledge of carbonate ore. A tunnel 100 feet in length has been driven on the vein. The Idaho shows a vein 4 feet wide, some of which assays as high as $200. It is opened by a shaft, 100 feet deep. The Susana is a vein of carbonate ore, going $60 per ton. A shaft 50 feet deep has been sunk on the property. The California has a shaft 100 feet deep. It is a strong vein of carbonates, that assays $50 per ton. Among the prospects worthy of mention in this district should be named the Sonora, North Carolina, Franco-American, Ready Relief, Southern Pacific, and many others.

EMPIRE DISTRICT.—This district is about twenty miles east of Tucson, in the rolling hills of the Rincon mountains. It is a short distance south of the Southern Pacific railroad. The camp has been brought into notice by the discovery of the "Total Wreck," an immense body of chloride ore, over 50 feet in width, and assaying from $10 to $500 per ton. The ore carries silver and gold. It has the appearance of a contact vein, between porphyry and lime. Work is prosecuted steadily. Three thousand tons of ore are on the dump, and reduction works will be erected at once. The Champion is a 20-foot vein, with a shaft 50 feet deep. The Dividend, Cross, Crescent, Ophir, and many others are on the same vein as the Total Wreck. They show large bodies of the same character of ore, and promise to become valuable properties.

OLD HAT DISTRICT is on the northern end of the Santa Catarina range, and thirty-five miles from Tucson. It contains plenty of wood and water, and is well situated for mining. The Bonanza has two tunnels, 300 feet in length. It is a large vein, assaying from $50 to $100 per ton. Work is carried on steadily, and a fine property is being opened up. The Braganza is a strong vein, producing ore that goes from $50 to $200 per ton. The other prominent mines are the Old Hat, Bandit, American Flag, Palmetto, Pioneer, Morning Star, Black Bear, Silver Glance, Montezuma, Mermaid, Pilot, Lookout, Manzana, and many more. With its beautiful situation, and abundance of wood and water, this district is destined to become a prominent mining center of Southern Arizona.

SILVER HILL DISTRICT.—This camp is fifty miles north-west of Tucson, and only eighteen miles distant from the railroad. The Abbie Waterman is the leading mine of the district. It shows a body of carbonate ore nearly 10 feet wide. It is a fine smelting ore, and assays high in silver. The mine is opened by several shafts and open cuts, showing the same body of mineral from end to end of the claim. This promises to become one of the most valuable discoveries in Pima county. The Amelia is the north extension of the Waterman. It is a large vein of fine ore. The Mamie Griffith, Monarch, Government, Lancer, Little Joker, White Cliff, and Rodrigues' Purse are all large veins, carrying ore of a good grade.

PAPAGO DISTRICT lies to the south-west of Tucson. It embraces a large area of country known as the Papagueria. This region contains veins of gold, silver, and copper. Water can be obtained by sinking, and mesquite and palo verde wood is met with nearly everywhere. The Montezuma mine is in this region, and also the famous Cabibi mines, which are rich in silver and copper. The Pichaco mine, in this district, has been worked for many years, and has produced a large amount of bullion. The San Pedro, Cabriza, El Cantavo, and many other large and promising veins are in this portion of Pima county. Westward from this group are the Ortega mines, rich in copper and silver; and still farther west are the noted Ajo copper mines, which were worked extensively in early times, and the ore shipped from Port Libertad to San Francisco. All this portion of Pima is rich in mineral, and will yet become the seat of a prosperous mining industry.

AMOLE DISTRICT is west of Tucson and contains several valuable mines that assay from $100 to $1,500 per ton. The Cymbeline, the Homestake, and the Hope are all fine properties. The Neuguilla mine has a shaft 90 feet deep, showing a vein between 4 and 5 feet wide.

PIMA DISTRICT lies about thirty miles south-west of Tucson, in the low hills of the Sierritas. It has yielded ore of a high grade, and promises, with development, to become an important camp. The Esperanza and the Rough and Ready are the leading mines of the camp. The latter has produced ore going $700 per ton.

HELVETIA DISTRICT is situated on the eastern slope of the Santa Ritas. It has abundance of wood and water. It contains rich placer mines which have produced several hundred thousand dollars. The district has also some valuable veins of silver and gold, though but little work has yet been done.

Copper.—Pima county, besides its ledges of gold and silver, is also rich in copper. High-grade copper ores are found on the northern end of the Santa Rita range, about twenty-five miles south from Tucson. The outcroppings cover several hundred acres, and are composed of carbonates, red oxides, and copper glance. Some of the veins are nearly 50 feet in width, going from 15 to 25 per cent. The copper deposits in the Silver Bell district, fifty miles west from Tucson, are among the largest and most valuable in the Territory. They are immense dikes, in places 50 feet wide, carrying carbonates, and red and black oxides. A smelter, with a capacity of 30 tons, is being erected on this property by the Huachaca Mining Company. Besides the copper mines here alluded to, the whole region west of Tucson, to the boundary of Yuma, and south to Sonora, is rich in this metal.

YAVAPAI COUNTY,

The largest political division of the Territory, has long borne an enviable reputation for the richness and extent of its mines, and for years was the leading bullion producer of the Territory. The principal mineral belt of the county lies between the thirty-fourth and thirty-fifth parallels of latitude, and extends from the Apache line to the boundary of Mohave. There is no part of the Territory so generally blessed with those two important factors in mining operations, wood and water. The formation of the mineral-bearing portion of Yavapai county is mostly a granite; porphyry, slate, and quartzite are encountered in many places, while the northern part shows sandstone, trap, and rock of volcanic origin. The mineral veins are noted for smooth, well-defined walls, high-grade ores, and great variety of mineral combinations. In its production of gold, Yavapai is the leading county of the Territory. The metal is found in nearly every portion of the mineral belt, in alluvial deposits, and in ledges. Silver occurs in native, wires and nuggets, chlorides, horn silver, silver glance, ruby silver, sulphides, black sulphurets, and many other rich varieties. Copper is found in oxides, native, malachite, blue carbonates, and as grey copper.

The first mining by Americans in Yavapai county began in 1863, with the discovery of the rich placers at Weaver creek. About the same time the Walker party, from New Mexico, found the diggings of the Hassayampa and Lynx creek. Since then mining has been carried on with generally satisfactory results. Until the opening of the Southern Pacific railroad, two thirds of the bullion shipped from the Territory was produced in this county; and nearly half the mining locations in Arizona were made within its borders. Mining operations conducted by ignorant, incompetent, and sometimes dishonest men, have greatly retarded the development of Yavapai. Unfortunately, mining litigation has also done its share in this direction. But despite these obstacles, the intrinsic merit of the mines has been proven, and against bad management and costly litigation they have been made to pay. The opening of the railroad on the thirty-fifth parallel will give the mines of Yavapai all the advantages of cheap and rapid transportation; will bring its vast mineral wealth before the world, and make it, what its unrivaled climate and great natural advantages destined it to be, one of the leading mining camps on the coast.

PECK DISTRICT.—This district is thirty miles south-east from Prescott, in the northern foothills of the Bradshaw range. It was organized in 1875, and has become famous for the wonderful richness of its ores. It has every advantage in the way of wood and water. Owing to continuous litigation, the mines of the district have not been worked as mines of their richness and extent ought to be. No camp in the Territory has produced the same amount of bullion, considering the length of time it has been worked and the number of men employed.

The Peck is one of the leading mines of the Territory. Discovered in 1875, it was worked successfully till 1878, when the owners became involved in a lawsuit which has not yet ended. The mine produced during that short period $1,200,000. Ore worth from $5,000 to $20,000 per ton was frequently met with. Pending the settlement of lawsuits, one of the finest properties in the Territory is lying idle. The rich vein is about 18 inches wide, composed mainly of chlorides and carbonates. The average working test has been near $200 per ton. The mine is opened by a 400-foot shaft, and by four levels, aggregating 1,300 feet. A complete ten-stamp mill and roaster have been erected on the property. The Peck is a strong vein, with prominent quartzite croppings traceable across the country for several miles. The Silver Prince is situated about half a mile east of the Peck. It has produced very rich ore, similar in character to that of the Peck. Several tons of this ore shipped to San Francisco have averaged $1,000 per ton. The mine is opened by several shafts and drifts, and by a tunnel nearly 400 feet in length. The vein is a strong and well-defined one, the richer ore bodies occurring in chambers or bunches. The Black Warrior lies south from the Prince and on the same vein. It is one of the finest properties in the camp; has been thoroughly opened by shafts, drifts, and tunnels. It shows a vein from 2 to 3 feet wide, composed of sulphurets, galena, native and antimonial silver, assaying on an average, $200 per ton. The Warrior and the Prince are the property of a New York company. The Asa Buffum is a northern extension of the Peck. It shows a small vein of exceedingly rich chloride and carbonate ore that assays $1,000 per ton. The Alta is situated south from the Peck, and between that ledge and the Silver Prince. It shows a vein from 1 to 2 feet wide of high-grade chloride ore, giving an assay of $300 per ton. The Evening Star is the south extension of the Alta. It is opened by several shafts and drifts showing ore similar to the Alta, assaying from $300 to $1,000 per ton. The Lone Juniper is a south extension of the Black Warrior. It carries a vein from 1 to 2 feet of carbonates and chlorides assaying from $80 to $500 per ton. The property has been developed by several shafts, drifts, etc. The Doyle is south of the Warrior. It has a vein 2 feet wide of sulphuret ore. Average assays go from $50 to $100 per ton. The mine has a 100-foot shaft and 100 tons of ore on the dump.

The May Bean is the first south extension of the Peck, and has produced very rich ore. It is owned by the May Bean Mining Company, and is explored by several tunnels and shafts. The Curtin is the north extension of the Prince. It is a large vein, having but little work done upon it. The Silver Chief is situated between the Peck and the Silver Prince. It has a shaft 40 feet and shows rich ore. The St. Paul, some distance south of the Peck, is a large vein carrying ore that averages about $30 per ton. But little work has been done on it. The Austin, south of the St. Paul, has produced ore worth $5,000 per ton. There are a great many claims on the Peck ledge which show good surface indications. The most prominent of these is the General Kautz, opened by a tunnel over 100 feet in length. The New York is north of the Curtin. It is a large ledge, showing good ore and opened by several shafts.

TIGER DISTRICT.—This district is situated about thirty-five miles south-east of Prescott, on the southern slope of the Bradshaw range. No camp in the Territory has better natural advantages for the mining and working of ores. Wood is found in every direction, and water is abundant; while the climate is all that could be desired. The formation is a granite. The district was organized in 1871, and contains.

4

many large and regular veins of gold and silver. The cost of bringing in machinery, and the curse of litigation, have been the causes which have hindered the development of its valuable properties. The Tiger, which has given its name to the district, is one of the largest veins in the Territory. being over 70 feet between· smooth and compact walls. It was the first silver mine of importance discovered in northern Arizona, and has produced some of the richest ore ever found in the Territory. The mine is opened by a three-compartment working shaft, 300 feet deep, supplied with steam hoisting-works, and equipped with cages and pumps. The ore is a sulphuret, carrying native silver, and has worked on an average, $110 per ton. A ten-stamp mill with roaster attached has been erected. The mine has produced $200,000, $5,000 being gold. The claim is 1,200 feet long by 200 feet wide. The Hammond and Riggs claim is the second south extension of the Tiger.. It is a large vein, showing fine ore, and has a tunnel driven along the ledge 150 feet, and a shaft 65 feet deep. The Linn ground is the first extension north. It shows a strong vein of high-grade sulphuret ore, and is opened by several shafts and cuts. The Tiger is a true fissure and is located for three miles, the claims varying from 200 to 1,200 feet in length, most of which have been patented. Nearly all of the claims are opened by shafts and tunnels, showing large ore bodies similar in character to the discovery location.

The Gray Eagle is about two miles east of the Tiger. It is a large vein of sulphuret ore, carrying gold and silver. Average assays give $46 in silver and $22 in gold. It is opened by 350 feet of tunnels. The Oro Bonito lies between the Tiger and the Gray Eagle; it shows a 3-foot vein of gold quartz, some of which has worked $80 per ton, in arrastras. The mine is opened by several shafts and tunnels. The Eclipse is about two miles east of the Tiger; it has from 1 to 3 feet of chloride and horn silver ore, assays from which have gone up into the thousands. A 60-foot shaft has been sunk on the mine. The Lorena is a small ledge east of the Eclipse; the ore is a chloride of silver and goes about $200 per ton. There is a shaft 80 feet deep and 100 tons of ore on the dump. The California and Benton are supposed to be northern extensions of the Tiger. They are strong veins and carry high-grade ore; the former has a shaft 100 feet, and the latter 50 feet deep. The Moreland is the north extension of the Benton; it is a large vein, carrying some very rich silver ore.

The Buckeye is situated in what is known as Bradshaw Basin. It is a small ledge of very rich ore—gold and silver. It has produced several thousand dollars, and is opened by shafts and tunnels. The Kansas is east of the Buckeye; it has a strong vein of sulphuret ore, and has been explored by a tunnel, over 100 feet in length. The Thurman is a 3-foot vein of sulphurets carrying gold and silver, and assaying $60 per ton. Several shafts have been sunk on the property. There are many other valuable claims in the "Basin" on which but little work has been done. A ten-stamp custom mill has been erected at this place.

North of the Tiger district, in what was formerly known as Pine Grove, are several fine properties, foremost among which is the War Eagle, a vein from 2 to 5 feet wide, carrying gold and silver, which has worked from $25 to $40 per ton. The discovery claim is opened by a shaft 90 feet deep. It has produced over $30,000, and is one of the most valuable properties in the Bradshaw. The claim has been located for several miles, the extensions all showing finely. The Del Pasco, Bradshaw, Blandena, Cougar, Gretna, Shelton, and many other promising claims, are in the Tiger and Pine Grove districts. About five miles west of the Tiger is located the Southern Belle, a ledge of gold quartz from 4 to 5 feet wide; the ore, worked in arrastras, has yielded from $30 to $50 per ton. Several shafts and open cuts show a well defined lode.

TIP TOP.—This district is about fifty miles south-east of Prescott in the spurs of the Bradshaw range. The camp has long been noted for the richness of its ores, and is a favorite of "chloriders," or poor miners who get out their "rock" and have it reduced at custom mills. The formation is a micaceons granite, and the veins, though small, are compact and regular. The district has produced more bullion than any other in Yavapai county, and its mines steadily improve in size and richness as depth is reached. The Tip Top is the principal mine of the camp; it was discovered in 1875, and has been worked continuously ever since. The main working shaft is down nearly 600 feet, and the claim is thoroughly opened by levels, winzes, tunnels, etc. The vein averages from 1 foot to 18 inches in width; the ore is a sulphuret, carrying quantities of ruby silver, and assaying $300 per ton. A 10-stamp mill and roaster is in operation on the Agua Fria, about nine miles from the mine. This is one of the best properties in the county, and has produced over $1,200,000. The Cross-cut is west of the Tip Top, and is the largest vein in the district. It is traceable across the country for several miles, and located nearly all the way. The Foy, a location on this ledge, shows 2 feet of ore assaying from $75 to $200 per ton. It is opened by a shaft 180 feet deep, and by several open cuts.

The Pearl, another location on the Cross-cut, is opened by a shaft 100 feet deep; it shows a strong vein of high grade milling ore, and is one of the most promising claims in the camp. The Swilling is north of the Tip Top; it has two shafts, 110 and 50 feet, respectively. It carries a 3-foot vein of milling ore assaying $50 per ton. The Virginia No. 2 is on Tula creek, about four miles from Tip Top. It shows 18 inches of free-milling ore, ranging by assay from $100 to $1000 per ton. The mine is opened by two shafts, 140 and 80 feet deep, and has produced $10 000 silver. What is known as the Rowe claim is near the Cross-cut; it contains some very rich ore, and is opened by a tunnel and several shafts. A number of tons of ore from this mine have been shipped to San.

Francisco, averaging from $500 to $1000 per ton. The Basin mine is three miles west of the Tip Top. It has been worked for several years—the ore being reduced in a custom mill—and has paid a handsome profit to its owners. The mine has produced a great deal of bullion, but the exact figures are not at hand. The "76" has a small vein of high grade ore of a similar character to the Tip Top, which assays from $200 to $1000 per ton. Three tunnels—200, 120, and 85 feet each—have been driven on the claim. The Incas is a narrow vein of exceedingly rich ore, assaying from $100 to $1,800 per ton. These are only a few of the claims of this district; there are scores of others which carry rich ore and give every promise of becoming valuable when developed. The ores of the camp are nearly all silver-bearing.

HASSAYAMPA DISTRICT.—This district is situated about ten miles south of Prescott, in the midst of a heavily timbered and well-watered region. The Hassayampa creek, after which the district takes its name, has been worked for gold ever since the settlement of Northern Arizona, and has produced a great deal of money. The character of the ores in the Hassayampa region is a gold quartz on the surface, which gradually passes into silver as depth is reached. The formation is generally a granite, with some slate and porphyry. The Senator shows more development than any mine in the camp. It has been worked extensively, and has a shaft 200 feet deep, with levels, drifts, cross-cuts, etc. The vein is from 2 to 4 feet wide—iron, copper, and lead sulphurets, which have yielded from $25 to $40 per ton. The mine has produced $160,000 in gold. It has a ten-stamp mill. The Davis is about four miles south of the Senator, on Slate creek, a tributary of the Hassayampa. It is a large vein of sulphuret ore, averaging 5 feet wide. It is opened by a tunnel nearly 100 feet in length. The ore gives an assay of from $50 to $300 per ton. The Davis is traceable across the country for nearly two miles, and several extensions, showing good ore, have been located on it.

The Crook is three miles east of the Hassayampa. Some of the richest gold quartz ever taken out in the county came from this mine. It is opened by 670 feet of shafts and 850 feet of tunnels. It has a vein from 1 to 4 feet wide, yielding $28 per ton. The claim has produced over $50,000 in gold, and shows good ore in every drift and stope. It has a ten-stamp mill. The Perry is eight miles south of Prescott. It is a strong vein of sulphuret ore; has a shaft 75 feet and a tunnel 185 feet. Selected ore from this mine has yielded $400 per ton, in silver. The Pine Tree shows a vein 18 inches wide, giving an assay of $90 per ton. It carries silver and gold, and is opened by a tunnel 350 feet in length. The Savage has two shafts, 40 and 50 feet. It carries 18 inches of ore, worth $200 per ton. The Cash has 2 feet of base ore, assaying $60 per ton. It has a shaft 28 feet. The Consolidated Bodie shows 4 feet of galena and carbonate, assaying $60 per ton. It has two shafts, 100 and 180 feet. The claim is on the east fork of the Hassayampa. The Sumner is a large vein, 45 feet of micaceous iron, portions of which are rich in silver. Assays as high as $2,000 have been made from this mine. The Caney shows 2 feet of gold quartz that has worked $38 per ton. The Grovanor has a 3-foot vein of gold and silver ore that assays $27 per ton. It is opened by three shafts. The Madison shows 1 foot of black sulphurets of silver worth $200 per ton. The Delaware has a 50-foot shaft, with 2 feet of ore, carrying gold and silver. The Montgomery, Peacock, Often, Railroad, Nevada, Connecticut, Constantine, Glenn, Fox, Kautz, Silver Trail, Jim Davis, United States, Storm Cloud, Fremont, Sterling, Nifty, and hundreds of others, are located in the Hassayampa district. It is impossible in this brief sketch to give more than a passing glance at the many valuable properties in this camp, the most delightful summer resort in the Territory.

WALKER DISTRICT.—This district is about seven miles east of Prescott, and embraces the headwaters of Lynx creek, the richest gold-bearing stream yet discovered in the Territory. It is estimated that over $1,000,000 has been taken from this creek since its discovery in 1863. Lynx creek is blessed with an abundant supply of wood and water, and a delightful climate. The veins carry gold and silver. The Shelton is a 4-foot ledge of carbonate ore, impregnated with iron pyrites. Assays go as high as $600 per ton. The ledge has a shaft 30 feet and a tunnel 100 feet. The Pine Mountain is a 2-foot vein of carbonate ore, assaying $120 per ton. It has a shaft 20 feet. The Gray Eagle has a tunnel 70 feet. It carries 4 feet of carbonate and sulphuret ore, assaying $80 in gold and silver. It has produced $4,000. The Mount Vernon carries 10 inches of rich gold quartz, worth $200 per ton. It has produced $15,-000, is opened by several shafts, and a tunnel 100 feet in length. The American Flag shows 18 inches of base-metal ore that has worked $50 per ton. It has a 50-foot shaft, and has yielded $3,000. The Hidden Treasure shows a ledge 12 feet wide, assaying from $27 to $200 per ton, gold and silver. It is opened by a shaft 50 feet deep.

The Accidental is thoroughly opened by shafts, tunnels, drifts, etc. It is a rich vein, carrying gold and silver, and has produced over $50,000. The Mountain Lion is a promising-looking claim, carrying gold and silver. It has a tunnel 135 feet in length. The Orion is a 4-foot vein of sulphuret ore, assaying $ 0 per ton. A shaft 42 feet deep has been sunk on the ledge. The Hirshel has a 6-foot vein of carbonate and galena ore, assaying $50 per ton in silver, and $15 in gold. It has a shaft 65 feet deep. The Capital is opened by a tunnel and several open cuts. It carries 2 feet of chlorides, assaying $80 per ton. The Real del Monte, Empire, Mark Twain, Champion, Henry Clay, Pointer, Boston, Eureka, Eberhardt, Alturas, and scores of other valuable locations, are in this district.

TURKEY CREEK.—This district is about twenty miles southeast of Prescott. The ledges are principally silver-bearing, in a

granite and porphyry formation. The camp has plenty of wood
and water, and a climate unsurpassed in the Territory. The Tus-
cumbia is situated on a northern spur of the Bradshaw range.
It is opened by a shaft 100 feet deep, and a tunnel 250 feet.
The vein is about 18 inches wide, assaying $200 per ton. A
five-stamp mill has been erected, and some $25,000 in silver has
already been taken out. The Goodwin is from 4 to 6 feet, between smooth walls; the pay streak is from 1 to 3 feet, assaying
from $50 to $800 per ton. The Holmes claim, on the Goodwin ledge, shows a vein from 1 to 4 feet wide, assaying from
$60 to $1,000 per ton. The ore is a rich antimonial silver. There
is a shaft 85 feet, and a tunnel 160 feet. It has yielded $2,800,
in silver. The Hatz and Collier claim is a northern extension of
the Goodwin. It shows a vein from 2 to 4 feet, that assays
from $50 to $500 per ton. It is opened by several shafts and
tunnels. The Continental is a large ledge, carrying a rich pay
streak. It has a tunnel 200 feet in length, besides several
shafts. The Peerless is a large dike, with a vein of rich ore.
It has a shaft 90 feet. The Succor shows a 2-foot vein of galena ore, some of which assays $200 per ton. It has a shaft 100
feet. The Gold Note shows a good body of rich galena ore. It
is opened by a shaft 100 feet deep. The Morning Glory is a
large ledge of gold quartz. It is opened by a shaft 100 feet
deep and by a 100-foot tunnel. The Trinity carries from 6 inches
to 15 inches of sulphuret ore, that assays $100 per ton. It has
a shaft 80 feet. The Compton has a shaft 48 feet, and carries
from 1 to 2 feet of antimonial silver ore. The Bully Bueno,
Town Site, Adirondack, Lincoln, Nevada, McLeod mine, Richmond, Kendall, Franklin, and many more, all show ore of a
high grade.

Big Bug.—This district is situated east of Lynx creek, and
about twelve miles from Prescott. It is surrounded by a forest
of pine timber, and has abundance of water. The ores carry
gold and silver. Considerable placer gold has been taken from
this camp. The Bell has three feet of argentiferous galena ore,
assaying $80 per ton. It is opened by a shaft 260 feet deep,
and by a tunnel 200 feet long. It carries gold and silver. The
Plat Bonita has a shaft 70 feet. It carries 4 feet of milling ore,
assaying $60 per ton. It contains silver and gold. The Mid-
dleton shows 4 feet of milling ore, assaying $50 per ton. It is
opened by a shaft 70 feet deep. The Poland has a tunnel
60 feet. It has a 3-foot vein of smelting ore, assaying $50
per ton. The Dividend is a 3-foot vein of gold-bearing quartz
that has worked $20 per ton. It has a shaft 120 feet. The Ga-
lena is a ledge of gold quartz that has yielded $20 per ton. It
has three shafts, 80, 100, and 125 feet, each. The Big Bug
shows 3 feet of base ore that has worked $20 per ton, in
gold. It is opened by an 80-foot shaft. The Eugenia shows
$2\frac{1}{2}$ feet of gold pyrites. It has a tunnel 100 feet in length.
The Belcher is opened by several shafts and tunnels. It carries $2\frac{1}{2}$ feet of free-milling gold ore that has yielded $20 per
ton. The Lottie is a 4-foot vein of milling ore, carrying gold

and silver, and assaying $60 per ton. It has a 100-foot shaft. The Champion has a shaft 100 feet deep. It is a 4-foot ledge of milling ore, assaying $60 per ton, and containing gold and silver. The Mesa, Pine Tree, Forest City, Black Fox, Challenge, Oury, Belcher, Ticonderoga, Independence, Crown Point, Rebel, Bunker Hill, Hamilton, and many other valuable prospects, are in.Big Bug. Several Eastern companies are now operating in the district.

GROOM CREEK.—This camp is about six miles from Prescott, in one of the finest timbered and watered regions of Northern Arizona. The ledges are regular and compact in a granite formation; they carry gold and silver. The Lone Star has a vein 2 feet wide of argentiferous galena ore, assaying $100 per ton. It has two tunnels, 94 and 74 feet, respectively, besides several shafts. The Golden Chariot is a 2-foot vein of gold and silver ore. It is opened by several shafts and drifts. The Mountain shows a two-foot ledge of gold quartz. It has a 70-foot tunnel. The Dauphin has a shaft 45 feet. It shows a strong vein, 4 feet wide, of free-milling ore, carrying gold and silver. The Mirabile has 18 inches of high-grade free-milling ore. It is opened by a shaft 85 feet deep. The Minnehaha has two shafts, 35 feet each. It carries 20 inches of rich milling ore, some of which, shipped to San Francisco, has gone $300 per ton. The Nevada shows 20 inches of milling ore, assaying $160 per ton. It has two shafts, 40 and 35 feet each. The What Cheer is a large vein of free-milling ore, with a 50-foot shaft. Select ore from this mine has gone $300 per ton. The Alcyone shows 2 feet of galena that assays $60 per ton. A 35-foot shaft has been sunk on the claim. The Surprise has a shaft 40 feet and carries 2 feet of free-milling ore, assaying $150 in gold and silver. The Homestead, Uncle Joe, Adell, Heathen Chinee, Gazelle, Chicago, Old Put, Black Hawk, Canadian, Alta, Providence, Wakefield, Gray Eagle, Omaha, Benjamin, and scores of others well worthy of special mention, are in this district.

CHERRY CREEK.—This camp is situated about twenty-five miles east of Prescott, on the southern end of the Black Hill range. It is on the main road to the Verde; has plenty of wood and water, and a desirable situation. The ores carry gold and silver, and are easily reduced The Black Hills is a ledge of argentiferous galena ore, 12 feet wide, and assaying $40 per ton. The mine is opened by a 40-foot shaft. The Silver Streak is a 4-foot vein, assaying $50 per ton in gold and silver. It has a 25-foot shaft. The Rustic shows 18 inches of carbonate ore, that assays $150 per ton. It is opened by a 35-foot shaft. The Hiawatha has a shaft 30 feet deep, with 2 feet of carbonate ore; assaying $50 per ton. The Hercules is a strong vein, nearly 4 feet wide, carrying silver and copper. It assays $60 in silver. The Sarah Jane shows a 2-foot vein of gold quartz, that assays $75 per ton. It has a 35-foot shaft. The Gold Ring, Carbonate Chief and Parole are fine-looking prospects, carrying rich ore and good-sized veins. There are numerous other locations in this district deserving mention, which must be omitted for want of space. Considerable gold is being taken out with arrastras, and the claims are steadily improving as they are sunk upon.

WEAVER.—This district is the oldest in the county, having been organized in 1863, after the discovery of the rich gold deposits of Rich Hill. In a depression on the summit of this mountain about $500,000 in coarse gold was found lying on the shallow bed-rock, near the surface. The gulches running down from this mountain were also rich in placer gold; they have been worked since their discovery up to the present time, and have produced, it is estimated, $500,000, making the yield of Weaver district in placer gold, $1,000,000. The ores of the district are nearly all gold-bearing. Weaver is about thirty-five miles south from Prescott. The Leviathan is an immense gold-bearing quartz ledge, in some places 300 feet wide. Assorted rock from the mine has worked $50 per ton, in arrastras. It is opened by a tunnel, which cuts it 100 feet below the surface, and by several shafts and cuts. It is estimated there are 2,000,000 tons of ore in sight in this enormous vein. The Marcus shows a vein, 3 feet wide, of free-milling gold ore, to a depth of 68 feet, after which it changes to a sulphuret. The free-milling ore, worked in arrastras, has yielded as high as $200 per ton. The vein is opened by an incline 85 feet deep, and by a shaft 65 feet, connected by drifts. A new working shaft has been started, and is now down sixty feet. There are scores of other promising properties in this district, among which may be mentioned the Metallic Candle, with a shaft 40 feet deep, and a vein of gold quartz 20 feet wide; the Emerald has a tunnel 125 feet long and a shaft 20 feet deep; the Buckeye has a shaft 30 feet deep; the Cosmopolitan has a shaft 20 feet; the Sexton has a shaft 20 feet. Between thirty and fifty men are steadily at work in the placers of this district, all making good wages.

THE MARTINEZ DISTRICT joins Weaver on the west. The ledges are gold-bearing, large, and well defined. A mill has been erected on the Cumberland, but incompetency and mismanagement caused it to prove a failure. The mine carries ore that assays $250 per ton, and has yielded over $4,000 from arrastras. The Martinez is a 6-foot vein assaying heavily in silver. The Model is situated in People's valley, but is included in Weaver district. It is a small vein of rich gold quartz, which averages about $50 per ton. It is opened by a shaft 150 feet deep, and by several tunnels. A Huntington mill, with a capacity of 5 stamps, has been erected on the property. The Miner is near the Model. It shows a vein of gold quartz 2 feet wide. It has a 100-foot shaft and several drifts, cross-cuts, etc. A five-stamp mill has been put up on the claim and is working satisfactorily. The mine is owned by the Bedrock Mining Company.

TONTO BASIN.—This district is in the south-eastern part of Yavapai county. The veins are large, carry gold and silver, and there is a plentiful supply of wood and water. The House and Rouse claim has a shaft 100 feet and 250 feet of tunnels. The vein is 5 feet wide, free-milling silver ore. The Dougherty shows a 5-foot vein of silver ore, and is opened by a 94-foot shaft. The Osceola has a shaft 115 feet deep and a 6-foot ledge of gold quartz. The Golden Wonder is opened by a shaft 150 feet deep. The ledge is 5 feet wide. The Zulu is a large vein of free-milling gold and silver ore, assaying $100 per ton, and opened by a shaft 65 feet deep. The American, Silver Belt, Excursion, Last Chance, Accident, and many other valuable locations, have been made in this district.

SILVER MOUNTAIN.—This district is south of the Tiger in the foothills of the Bradshaw range. Some of the largest ledges in the Territory are found here. The formation is granite and porphyry. Among these immense veins may be mentioned the Mammoth, from 50 to 300 feet wide, and traceable across the country for several miles. Five claims of 1,500 feet each, have been located on the ledge. But little work has yet been done, although some very fine ore has been taken out. The mine contains silver and gold. Among the other veins of unusual size are the Excelsior, Great Western, Mountain King, and Snowball.

WALNUT GROVE.—This district is about thirty miles south of Prescott, and embraces the eastern end of the Antelope range. The veins are small, but rich in gold and silver. Wood and water are found in abundance on the Hassayampa. Among the prominent claims may be mentioned the Crescent, Josephine, Vesuvius, Rebel, and many others. A five-stamp mill has been erected in the district, but is now idle.

THUMB BUTTE.—This camp is six miles west of Prescott, in the Sierra Prieta range. It contains some small veins carrying very rich silver ore. The surroundings are all that could be desired; wood and water are found everywhere. There are several claims with shafts from 20 to 70 feet, all showing fine ore.

AGUA FRIA.—This district is sixteen miles east of Prescott, in the foothills west of the stream of the same name. The ore is silver and of a very high grade. The mines are in contact formation between slate and granite. The Silver Belt is the leading mine of the district. It is opened by three shafts, 65,110, and 165 feet in depth. The ore is a carbonate, carrying chlorides, horn silver, and native silver, and yields $250 per ton. The ore is smelted and the base bullion shipped to San Francisco. The capacity of the furnace, which is run by the water power of the Agua Fria, is 7 tons in 24 hours. The Belt has produced nearly $100,000 in silver. The Kit Carson, Silver Flake, Agua Fria, and Raible and Hatz claim, are the other principal mines in the district.

BLACK CANYON.—This district is twenty-five miles east of Prescott. It extends from the eastern spurs of the Bradshaw range to the Agua Fria. The veins are principally gold-bearing, with regular and well-defined walls. There is plenty of timber on the slopes of the Bradshaw range, and water at all seasons in the Agua Fria and the Black Canyon. This latter stream, which drains a large area of mountain country, has produced considerable quantities of placer gold, and is yet being worked for the metal. The leading claims of the district are the Ballenciana, the Wickenburg mine, the Zika mine, and the Sonora. The Zika claim is worked in arrastras run by water power on the Agua Fria. It is a strong vein of gold quartz. The Wickenburg has a large ledge of quartz, showing pure gold all through the vein. It is worked by arrastras.

Copper.—Yavapai is rich in copper ores; they are found in every part of the county, some of them of a very high grade. Very large deposits are found east of the Agua Fria and in the southern end of the Juniper range; copper is also found in the Walnut Grove district, in the country west of Date creek, and in Castle creek, south of the Bradshaw mountains. So far as developed, these deposits show ore of a high percentage, and of a character easily reduced. The only copper mines which have been thoroughly opened are situated in the Black Hills, about twenty miles north-east from Prescott. The Eureka, the leading mine of the group, has been explored by several tunnels, which have tapped the vein nearly 200 feet below the surface. The ledge is from 8 to 16 feet in width, and over 1,600 tons are on the dump. The property has recently been purchased by Eastern parties, who intend to erect reduction works. The Wade Hampton is on the same ledge as the Eureka, and carries a large ore body similar in character. There are many other promising claims in this district, which possess the advantages of wood and water, and will be only forty miles from the Atlantic and Pacific railroad.

PINAL COUNTY.

The first mineral discoveries were made in what is now Pinal county in the fall of 1871, but the hostility of the Indians and the isolated condition of the Territory at that time prevented any real development. The region was known to be rich in the precious metals, and after a peace was conquered from the Apaches, in 1874, prospectors flocked thither. The discovery of the famous Silver King in the fall of 1874, was the beginning of permanent mining in Pinal, and since that time it has been prosecuted without intermission and with the most flattering results. Few counties of the Territory can show a better record of bullion production. The Gila river, which flows through the center of this mineral belt, affords an unlimited water supply; while wood is found everywhere sufficient for all purposes of ore reduction. The ore bodies of Pinal county are noticeable for their size and richness. The formation of the country rock varies according to the locality, but granite and porphyry appear to be the predominating formations. A basaltic cropping is found in some places, and quartzite is sometimes met with. Gold, silver, and copper are the leading metals of Pinal. Large bodies of coal of an excellent quality, have lately been discovered in the eastern portion of the county. A railroad has been surveyed from Casa Grande to Pinal City, which will pierce the center of the mining region, and will be of incalculable benefit to the county. In its bullion product at the present time, Pinal takes the second place among the counties of the Territory; and with the opening of the projected rail communication, we may look to see that production largely increased. .

PIONEER DISTRICT.—The Silver King, the leading mine of Pinal, is situated in this district. The croppings of the vein are on a low, conical hill, in a basin, surrounded by spurs of the Pinal mountains. It is said that the mine was discovered from information furnished by a discharged soldier, who was stationed at this point during the Apache wars. After working the mine for nearly two years and taking thousands of dollars from surface excavations, the original locators sold the property to the present company, and the work of development was begun systematically. The discoverers of this magnificent property—farmers in the Gila valley—"builded better than they knew" when they conferred so appropriate a name on the wonderful mine. The vein matter is chiefly quartz; the ore is a sulphuret, carrying large quantities of native silver, polybasite, copper glance, blende, antimony, and other combinations. No such bodies of native silver have been found on the coast. The main working shaft is down over 600 feet, showing fine ore in the bottom; five levels have been run, and cross-cuts and winzes have thoroughly opened the mine. In places the ore body is 85 feet wide. A twenty-stamp mill has been put up at Pinal, five miles distant, and also a roaster and concentration works. The yield of bullion for May, 1881, amounted to $99,000. The ores are worked by the lixiviation process, which has proven a complete success. Whether we consider the size of the vein, the richness and variety of its ores, or its bullion yield, the King must be ranked as one of the great mines of the world.

The North King has a shaft 450 feet in depth, with improved hoisting machinery. The South King has been sunk to a depth of 150 feet. The property is owned by San Francisco parties. The Eastland is down 200 feet, and work is prosecuted steadily. This mine is producing some fine ore. Hoisting-works have been erected. The Last Chance shows a vein nearly 5 feet in width. The ore is a sulphuret, rich in silver. There is a tunnel on the claim 160 feet in length. The Mount View has a 4-foot vein of argentiferous galena ore. It is opened by a shaft 100 feet in depth. The Alice Bell shows 4 feet of galena ore, carrying considerable silver. It has a tunnel 80 feet.

The Belcher is one of the leading mines of the district. It is a chloride ore which gives an average of $82 per ton, the vein being from 3 to 5 feet wide. A ten-stamp mill has been erected on the property. The Eureka is on the same ledge as the Belcher. It shows a vein of chloride ore from 2 to 3 feet wide, assaying $100 per ton. It has produced about $5,000. The Surprise is a large gold ledge, 8 feet wide, assaying $40 per ton. It is owned by the Surprise Mining Company, who intend to erect a twenty-stamp mill the present year. The Gem is a 5-foot ledge of gold quartz. A mill has recently been erected on the property, which is owned by the Wide Awake Mining Company. The Lewis shows 4 feet of carbonate ore. It is on the same ledge as the Belcher and Eureka. The Silver Bell has 2½ feet of free-milling ore that assays $100 per ton. It has produced $10,000, and is owned by the Silver Bell Mining Company. The Silver Queen is the first location made in Pioneer district. It is a large body of metal, rich in copper and silver. The mine is opened by shafts, drifts, etc., and shows good ore throughout. Some distance south-west of Pinal is a group of mines lately sold to a California company. The ore in these claims is a carbonate, rich in silver. The veins are large, and well situated for wood and water. The company who purchased is known as the Pinole Mining Company.

MINERAL HILL.—This district is in the foothills of the Pinal mountains, about fifteen miles north-east of Florence. The formation of the district is granite. The Gila valley furnishes both wood and water. The ore is smelting, the veins large, and of a good grade. They carry gold and silver. The Alice shows a ledge from 6 to 10 feet wide, carbonates and galena. Assays from this vein give $80 per ton. There is a 60-foot shaft and 180-foot tunnel on the property. The Pacific is a ledge from 8 to 20 feet wide. It is opened by four shafts, the deepest being 60 feet. Ore from this mine has assayed $100 per ton. The Le Roy is a 6-foot vein, going from 40 to 50 ounces per ton. It has a shaft 100 feet, and a tunnel of 150 feet. The Chocia is an immense vein, from 6 to 30 feet in width, portions of it assaying 50 ounces per ton, silver. A shaft 50 feet has been sunk on the property. The lodes of this district offer many advantages for a successful mining enterprise, and a prosperous camp is certain to spring up here.

QUARJARTA.—Quarjarta district lies about six miles south of the Southern Pacific railroad, at Casa Grande. There is plenty of mesquite, palo verde, and ironwood, and water can be had by sinking in the bed of the Santa Cruz. The district has produced very rich ore, which was shipped to San Francisco, before the building of the railroad. The Quarjarta mine is opened by a shaft 146 feet deep, and by several open cuts. It shows 4 feet of ore that assays $60. The east extension is a large vein of carbonate ore. It has a shaft 50 feet, and 38 feet of drifts. It assays $50 per ton. The Antelope shows a 6-foot vein of gold quartz. Selected ore from this mine has yielded $100 per ton, in arrastras. There is a 40-foot shaft on the claim. The Sacaton is a fine-looking prospect. It shows a ledge over 14 feet wide, of carbonate ore, that assays $40 per ton. There are many other promising prospects in this district, which, with development, may prove valuable.

On SADDLE MOUNTAIN, between the junction of the San Pedro and the Gila, some discoveries of large carbonate veins have recently been made. The new camp is well situated, being in the center of a wooded region, while the Gila and the San

Pedro furnish an inexhaustible water supply. The Hayes, the most prominent of these new discoveries, has a vein of carbonates from 2 to 4 feet wide, which assay from $30 to $500 per ton. There is an 80-foot shaft on the claim. On the Golden Eagle there is a shaft 75 feet, and a cross-cut of 26 feet; the ore is carbonate, carries gold and silver, and assays from $70 to $100 per ton. The Maybell has a 60-foot shaft, and assays $100 per ton. There are some fifteen other locations, all showing good ore. A 5-stamp custom mill will soon be in operation at the mouth of the San Pedro.

RANDOLPH DISTRICT is situated in the Superstition range, north-west from Pinal City. The ledges of the district are large, with ore of a high-grade. The first discovery, known as the Randolph, is over 40 feet wide, the ledge being traceable across the country for several miles. The ores are mostly carbonates and chlorides of silver. Assays run all the way from $30 to $1,000 per ton. There is plenty of water, and wood can be had six miles distant.

CASA GRANDE. DISTRICT is situated about 20 miles south from the station of the same name on the Southern Pacific railroad. It is a late discovery, and the ores are said to be of a very high grade. Its proximity to the railroad gives this new district many advantages. A lively camp has sprung up about the mines, and the work of development is pushed forward vigorously.

Copper.—On Mineral creek, a tributary of the Gila, northeast from Florence, in the foothills of the Pinal mountains, are situated some rich copper mines. The camp is about five miles from the Gila river, and abundance of wood is found in the neighborhood. No better situation for a mining camp can be found in the Territory. A smelter with a capacity of 30 tons in twenty-four hours, has been erected at the river, and is running successfully. The Keystone is a large ledge carrying great quantities of native copper. The average of the ore is said to be about 25 per cent. The mine is opened by several shafts, drifts, etc. The Ida Ingalls is a 14-foot vein of copper glance, a large part of it giving assays of 30 per cent. There is a shaft 100 feet, and a drift 60 feet on the property. The Monitor shows 7 feet of good ore. It has several openings.

GILA COUNTY.

This county was created by act of the legislature of 1881, from portions of Pinal and Maricopa, and is one of the most thoroughly mineralized divisions of the Territory. Gold, silver, copper, lead, coal, and iron are found within its borders. In the richness of its silver ores, the region now embraced in Gila county has long been famous. With the exception of the Planchas de Plata, no such bodies of pure silver have been found in the Territory. This region was once the home of the Pinal Apaches, who guarded long and well the treasures which were known to be hidden in their mountain homes. As early as 1871, an expedition numbering nearly 300 men, and led by the Governor of the Territory, penetrated this region, but as their quest was for placer gold, they discovered none of the rich silver lodes over which they passed. It was not till 1875, that Globe district was organized and its boundaries defined. At that time the larger portion of the present county of Gila was embraced within the limits of the San Carlos Indian reservation, but the richness of the new discoveries caused the boundaries of the reservation to be narrowed, and the mineral region declared a portion of the public domain.

The geological formation of the county is generally granite, porphyry and syenite. Quartzite is found in several places, and also limestone and micaceous slate. The rolling hills adjacent to Pinal creek show large beds of cement overlying the primitive rock. Water in abundance is found by sinking in the washes and gulches throughout the county, while Pinal creek is a running stream for nearly nine months in the year, and carries at all seasons in its underground channel, water in abundance for the purposes of ore reduction. Of wood, it is estimated there are 40 square miles of pine in the Pinal mountains, besides oak and juniper in large quantities in different portions of the county. The ores of Gila show a great variety of mineral combinations. In the Pinal mountains they are a sulphuret, carrying base metal, and requiring to be roasted before being milled. In the vicinity of Globe, Richmond Basin, and McMillenville, the ores are generally free-milling, with some iron and copper. The copper ores of the county are generally of a high grade and easily reduced. Gila county has rich mines, and many of them; it has wood and water in plenty; its climate is unsurpassed; a railroad will soon tap the mineral field; capital is steadily seeking investment, and the future of this region is as bright as its past has been prosperous and productive.

GLOBE DISTRICT.—This district embraces the leading mines of Gila county. Probably no portion of the Territory of the same extent has produced ore of such wonderful richness. Tons of this ore, shipped to San Francisco in the early days of the district, have given the Globe country a reputation which has extended all over the coast. Among the leading mines of the district we enumerate the following: The Irene is a strong vein of carbonate ore, in some places 20 feet wide, and carrying a pay streak of about 6 feet, which will go close to $80 per ton. The mine is opened by a shaft 240 feet deep, and a tunnel of 330 feet connecting with the shaft. The mine has well-defined walls. It is owned by the Irene Mining Company of New York. Over 300 tons of ore are on the dumps and a mill will soon be erected. The Alice is a 4-foot vein of free-milling ore, assaying $100 per ton. It is opened by a shaft 235 feet deep, and a tunnel nearly 300 feet in length. The property is owned by the Globe Mining Company. The Centennial shows 2½ feet of grey carbonates, worth $100 per ton. It has a shaft 100 feet, and two drifts, one of 130 feet and one of 100 feet. The Democrat is opened by a shaft 33 feet deep and by a tunnel of 30 feet. It shows a 6-foot vein of free-milling ore running from $40 to $150 per ton. It is a strong vein, and one of the most promising properties in the district.

The Stonewall No. 1 is a large ledge with croppings in places 20 feet high. Three distinct ledges are traceable the entire length of the claim. The center vein carries large quantities of horn silver; the others are rich in carbonates. The average of the main vein is 50 ounces silver, per ton. There are two shafts, 50 and 100 feet deep, respectively. The ore body at the bottom of the deepest shaft is 12 feet wide, with good walls. The California shows a vein 5 feet wide, that assays from $40 to $100 per ton. A tunnel has been run on the claim nearly 200 feet, and several shafts sunk, the deepest being 50 feet. The mine is about four miles north of the town of Globe. The Miami is a well defined vein, carrying 3 feet of ore that has worked $70 per ton. Two shafts have been sunk, 80 and 90 feet respectively. There is a ten-stamp mill attached to the property, which has produced over $25,000. The Champion shows a 4-foot ledge of free-milling ore. It is opened by a main shaft 125 feet deep, and by several drifts and tunnels. A ten-stamp mill has been erected, and has produced a large amount of bullion.

The Golden Eagle is a large vein of free-milling ore. The mine is thoroughly opened by shafts, drifts, tunnels, etc. A ten-stamp mill reduces the ore. The bullion yield has already exceeded $80,000. The Julius is six miles from Globe. It has produced some exceedingly rich ore, and over $10,000 has been taken from it. Fifteen hundred pounds of ore from this mine, worked in San Francisco, yielded $5,000. The Rescue is one of the first discoveries in Globe district. Several tons of ore shipped to San Francisco went over $1,000 per ton. One ton yielded $3,000. The vein is 4 feet wide, chloride ore. There is a tunnel 84 feet and a shaft 80 feet. The Emeline has a shaft 50 feet. It shows a compact vein of free-milling ore, 18 inches wide, which will average $150 per ton. The Chromo is one of the oldest locations in the camp. The ledge is made up of numerous spar veins, from a mere thread to 45 feet in width. The ore is found in these veins, and assays from $5 to $100 per ton. The ore is a chloride.

The Centralia is one mile from Globe. It is in a limestone formation, showing many beautiful fossils. There is a tunnel on the property, of 110 feet. The ore shows a carbonate, impregnated with a sub-oxide of iron No extensive workings have been made, but the claim shows a fine prospect. It has produced ore that has worked $224 per ton. The Townsend is owned by the Townsend Mining Company. The vein is from 2 to 8 feet wide, gold ore. Small quantities worked by arrastra process, have given over $50 per ton. There is a tunnel on the property 150 feet in length. There are over 300 tons on the dumps. The company own a five-stamp mill. The Fame has a small vein, about 1 foot in width, of chloride and sulphuret ore that assays from $80 to $600 per ton. It has a shaft 50 feet. The Independence has a shaft 120 feet, and a drift from the bottom, 30 feet. The vein is from 3 to 5 feet in width. The ore is a chloride, carrying malleable silver, and assays as high as $600 per ton. Selected ore has worked 584 ounces, silver. The mine is eight miles from Globe. The Anna is a large vein upon which but little work has been done. It has produced ore worth $170 per ton. The Cox and Copeland claim, seven miles from Globe, has produced several thousand dollars in native silver. Among the many other valuable claims in the vicinity of Globe, and within a radius of ten miles of the town, are the Bluebird, Buckeye, McCormick, Turk, Andy Campbell, South-west, Alice, Dondona, Florence, Empire, Imperial, and scores of others.

RICHMOND BASIN.—The camp is situated on the western slope of the Apache mountains and about fourteen miles north of Globe. Wood and water are plentiful. The veins are strong and well defined. This camp is famous for the native silver nuggets which were found on the surface. It is estimated that over $80,000 in pure silver was picked up in this locality. The McMorris, the leading mine of the camp, is a vein nearly 8 feet wide. The ore is a native silver, silver glance, and bromide of silver. The main shaft is down 400 feet. An incline has been sunk 300 feet, and a tunnel driven 100 feet. There are three levels aggregating 700 feet. The mine has been one of the most productive in Gila county, and the yield up to date is estimated at $400,000. Steam hoisting-works have been erected and also a ten-stamp mill. The Silver Nugget takes its name from the "planchas" which were found within its limits on the surface. Some of these lumps of silver weighed five pounds. The ore of the Nugget, is free-milling. The vein is large, and is opened by two shafts, one of 160, and one of 100 feet, and a drift 180 feet in length. The ore is worked in a five-stamp mill. The East Richmond is a 9-foot vein, has produced very rich ore, and is opened by two shafts, 100 feet and 30 feet, respectively. The West Richmond is an extension of the McMorris. It shows a vein 8 feet wide. It has a shaft 96 feet, and one of 35 feet. The Dundee is a 4-foot vein that assays $60 per ton. It is a fine-looking prospect. La Plata has a ledge 7 feet in width. A shaft has been sunk 60 feet, and a tunnel run 120 feet. It is an extension of the McMorris, and has been sold for $60,000. The Cora, South Plata, Rifleman, Belle Boyd, and a great many others in this camp, show every indication of developing into valuable paying properties.

McMILLENVILLE.—This group of mines is situated about twenty miles north of Globe, about six miles east of Richmond Basin, and almost eleven miles south of Salt river. Nearly all the locations are on one immense fissure, traceable across the country for twelve miles. The country rock is porphyry and syenite. Wood and water are abundant. The Stonewall is the leading mine of the camp. It is a very large vein, impregnated with chlorides and native silver. A stratum running into the main

vein, and from 1 to 2 inches wide, is almost pure silver, and has
yielded many thousands of dollars. Steam hoisting-works
have been erected on the property. The main shaft is down
600 feet, and there are over 700 feet of drifts, etc. A 5-stamp
mill has been erected, and the total yield of silver is said to
be $300,000. - The Hannibal is the first extension north of
the Stonewall, the vein being of the same size and of a similar
character. It has a shaft 160 feet, and 200 feet of drifts
and cross-cuts. The Washington is the second north extension of the Stonewall. Its shaft is down 100 feet. The
Centennial and the Virginius are the third and fourth
extensions of the Stonewall, north. The former has a
shaft 25 feet, and the latter shows several feet of ore that
assays $80 per ton. The R. E. Lee is the first extension south
of the Stonewall. It is a large vein of free-milling ore;
has a shaft 90 feet, and a drift 75 feet. The Henry Clay and
the San Francisco are also on the same vein, but have little development. The Democrat and the Little Mack are on a spur
which runs into the Stonewall vein. They have both produced
very rich native silver ore, the total yield being estimated at
$85,000. The North Star is north-west from the Stonewall. It
shows a ledge from 3 to 8 feet wide, and has produced ore that
has gone as high as $1,000 per ton. The mine is opened by a
120-foot tunnel and a 60-foot shaft. The Concord and the Nevada are promising claims. The former is over 20 feet, and the
latter about 10 feet wide, with a tunnel 200 feet in length, and
shows ore going from $5 to $60 per ton.

About sixteen miles south from Globe, on the southern slope
of the Pinal mountains, is a group of mines which show large
veins and high-grade ore. They are surrounded by a fine body
of timber, and never-failing springs of water. The South
Pioneer is the most prominent mine in the group. It is a
3-foot vein of sulphuret, rich in native silver. Assays from this
ledge have gone as high as $20,000 per ton. Work is pushed
forward steadily, and hoisting machinery and reduction works
will soon be erected. The property is being opened by three
shafts, the deepest at this writing, being 80 feet. The Pioneer
is one of the finest-looking properties in Gila county. The
Great Republic shows a 2-foot vein, assaying $150 per ton.
The ore is the same character as the Pioneer. The mine has
a shaft 80 feet deep. The Missouri mine is also a fine property,
carrying a strong vein of sulphuret ore, with beautiful specimens of native copper.

Copper.—Gila county contains some of the finest copper
properties in the Territory. The Globe copper mine is about
one mile from the town to which it has given its name. It was
the first mine located in what is now Gila county. It is a large
vein, and has been taken up for several miles. The ore is a
high grade, carrying $25 in silver. The True Blue is one of
the most promising copper properties in the district. It is
opened by several shafts and tunnels, and shows 3 feet of
ore that gives an average of 30 per cent. A smelter of 30 tons

5

capacity is nearly completed on this property, which is situated about three miles from Globe. The O'Doherty is another large vein, carrying high-grade ore, and opened by a shaft 50 feet deep. About eight miles from Globe, at what is known as the Bloody Tanks, is another group of copper ledges, on which a smelter of 30 tons capacity is now being erected by a New York company. The Chicago, New York, Old Dominion, and Buffalo are owned by this company. The veins are large, and the ore is said to be of a high grade.

MOHAVE COUNTY.

Mohave is purely a mineral region. Its agricultural resources are confined to a strip of land along the Big Sandy, and to the valley of the Colorado. There are portions of the county which afford good grazing, but mining must be its main, and we had almost said, its only industry. Almost every mountain range within its borders is seamed with rich veins of gold, silver, and copper. The distance from supplies, the cost of freight, and the want of proper reduction works, have hitherto prevented the proper development of Mohave's vast mineral wealth. The building of the Atlantic and Pacific railroad, which will pass through the center of the mining region, assures for this county, so long isolated and neglected, a bright future. The silver ores of Mohave are mostly sulphurets, carrying native silver, ruby silver, silver glance, and other rich combinations. Chlorides are also found, and some rich argentiferous galena. The veins are nearly all inclosed by well-defined walls. Water and wood are abundant in nearly every locality. A band of prospectors entered Mohave county in 1858, and explored the mountain ranges near the Sacramento valley. It was not until 1863, however, that any real work was done; but the hostility of the Hulpai Indians, who killed many miners in their shafts, compelled the abandonment of the country. In 1871 and 1872 the first permanent improvements were made. Since then Mohave county has struggled against every obstacle and disadvantage which her remote situation naturally entailed. The lack of reduction works necessitated the shipping of the ores to San Francisco, at an enormous expense. Ores that would not go $500 per ton left no profit for the owner. Despite these drawbacks, the county has steadily advanced; the great richness of its mines has been proven conclusively, and they only await the benefits of cheap transportation to become steady bullion-producers.

Hualapai District.—This district is situated in the Cerbat range, about 35 miles from the Colorado river. The formation is granite and gneiss. Wood is plentiful, and water in sufficient quantities for milling purposes. The veins are of fair size, and the ore is of high grade. The Lone Star has been worked to a depth of 200 feet, and is opened by over 300 feet of levels. It shows a vein of rich ore over 18 inches in width that assays $150 per ton. The ore is concentrated and shipped to San Francisco. It is a sulphuret, carrying considerable base metal. This mine has produced over $60,000. It has steam hoisting-works. The Keystone has a shaft 260 feet, one of 150 feet, and over 400 feet of levels, drifts, winzes, etc. A five-stamp mill, with roaster, has been erected on the property, and also steam hoisting machinery. The mine shows a 2-foot vein of sulphuret ore, that has worked $100 per ton. The property is owned by the New York Mining and Milling Company, and has produced over $100,000. The Fairfield is a 5-foot vein that assays $60 per ton. It has a shaft 185 feet. A tunnel is being pushed to strike the vein, which is now in 1,000 feet. The Stark and Ewing is an extension of the Keystone. It shows a 5-foot vein, and has a shaft 40 feet. The Ithaca has a vein ranging from 1 to 2 feet of chloride ore, assaying $70 per ton. It is opened by 500 feet of shafts, drifts, and tunnels. It has produced about $12,000. The Rattlesnake has a 70-foot shaft and a 75-foot tunnel. It shows 3 feet of chloride ore worth $50 per ton. All these claims are in the immediate vicinity of Mineral Park. About four miles north is the camp of Chloride, which shows rich ores and large veins, among which may be mentioned the following: The Connor, a 3-foot ledge, assaying $100 per ton. It has a shaft 100 feet, carries both gold and silver, and has produced $20,000. It is owned by the Arizona Northern Mining Company. The Empire has a vein of rich sulphurets, and has turned out $10,000. The Schuylkill is a 3-foot vein of carbonate ore, assaying from $50 to $75 per ton. It is opened by two shafts, one 65, and the other 40 feet. The Schenectady has a shaft 80 feet, with a 3-foot vein, running from $50 to $70 per ton. The Valley View is a large vein, running from 8 to 10 feet wide, with an average of $40 per ton. It is opened by three shafts and one tunnel. The San Antonio has a shaft 50 feet; a vein 2½ feet wide of free-milling ore, worth $100 per ton. The Donohue and the Rogers are also fine properties, and have produced about $18,000 each. In Todd's basin, about four miles south of Mineral Park, there is a group of mines which have considerable work done upon them. The most prominent are the Todd, a 4-foot vein of sulphuret ore, going about $60 per ton. The Oro Plata has a tunnel 100 feet, and several drifts and shafts. It has a 4-foot vein of free-milling ore, that goes over $50 per ton. It has produced $30,000. The Mariposa shows 18 inches of chloride ore, worth $150 per ton. It has two shafts, 40 and 30 feet, and has yielded $10,000. The Paymaster is a vein 3½ feet wide, assaying $60 per ton. A shaft has been sunk 50 feet. The mine has yielded $15,000. The Silver has a shaft 35 feet, and a body of ore 3 feet wide that averages $80 per ton.

CERBAT is about seven miles south of Mineral Park, in the mountain range of the same name. The country formation is granite. Wood is abundant, and water in quantities sufficient for ore reduction. The ores are generally of a high grade, but most of them carry sulphurets and require roasting before being milled. The Cerbat claim has a 4-foot vein that assays $100 per ton.

The ore is a sulphuret, carrying horn silver. It is opened by a shaft of 120 feet, and by two drifts, 80 and 65 feet. The ore carries gold and silver. A complete five-stamp mill with a roaster attached, has lately been erected. The mine has produced $25,000 from steam arrastras. The property is owned by the Arizona Northern Mining Company. The Fontenoy shows a vein 2½ feet wide that assays $125 per ton. The ore is a chloride of silver, and the mine has already yielded over $30,000. It is opened by two shafts, 110 and 75 feet, respectively, and a tunnel 65 feet. The Seventy-eight, formerly known as the Sixty-three, carries a vein of chlorides from 1 to 3 feet wide. Ore from this mine, shipped to San Francisco, has yielded, on an average, $350 per ton. The mine has produced in the neighborhood of $300,000. The claim is opened by 300 feet of shafts and 700 feet of tunnels. The Silver shows a vein from 2 to 4 feet wide, that assays $80 per ton. It has a shaft 80 feet and 50 feet of tunnels. The Flagstaff is opened by two shafts, one of 150 feet, and another of 100 feet. It has a vein 3½ feet wide that averages, by assay, $70 per ton. The Gold Bar has a 3-foot vein of gold quartz and a shaft 200 feet deep. Ores from this mine have been worked by arrastra process with satisfactory results. The Black and Tan is opened by a tunnel 250 feet in length, and a shaft 75 feet. It shows 2 feet of ore that assays about $100 per ton, and has yielded $20,000. The Flores has a vein of free-milling ore 3 feet wide, working $50 in gold, and $20 in silver. It is opened by a shaft 95 feet deep and a tunnel 200 feet in length. It has produced nearly $35,000, the ore being worked in arrastras. The Vanderbilt has one shaft 90 feet and another 50 feet. It carries gold and silver. The vein is about 2 feet wide, and the ore assays $70 per ton. The Tulare has 4 feet of ore that assays $50 per ton. It is opened by a shaft 110 feet, and has produced over $10,000 in gold and silver. The Bay State is a carbonate ore. Its vein is 4 feet, assaying $40 per ton. It has 3 shafts, 90, 80, and 30 feet. The New London shows 3 feet of galena ore, worth $50 per ton. Its deepest shaft is 95 feet, and it has produced $9,000 in silver. There are many other claims in the Cerbat camp showing good ore and large veins. Nearly all the bullion produced has been taken from ore shipped to San Francisco by poor mine-owners, who have made their claims pay against every disadvantage.

STOCKTON CAMP is situated on the eastern slope of the Cerbat range, about six miles south-east from Mineral Park. It has a delightful situation, fronting on the Hualapai valley, and is only eight miles from the surveyed line of the Atlantic and Pacific railroad. The formation is granite; wood and water are found in abundance. The camp has been self-sustaining, having received no aid from outside capital. The Cupel has produced about $150,000. It is a 2-foot vein, and has worked $100 per ton. It is opened by 500 feet of shafts and drifts. The ore is a sulphuret of silver. The Prince George shows a 3-foot vein that assays $80 per ton. It has 100 feet of shafts and has produced $12,000. The IXL has two shafts, 110 and 80 feet. The width of the vein is $3\frac{1}{2}$ feet, which assays $80 per ton. The Infallible is a strong vein 4 feet wide, with ore that averages by assay $70 per ton. It is opened by five shafts and 100 feet of drifts. It carries gold and silver, and has produced over $5,000. The Tigress has 18 inches of rich galena ore, worth $150 per ton. It is opened by several shafts and drifts, and has yielded $25,000. The Little Chief is a small but exceedingly rich vein. It has nearly 200 feet of shafts and other openings. The ore shipped has gone from $400 to $1,200 per ton. The total yield has been about $50,000. The Cincinnati, Bullion, Silver Monster, Fountain Head, Miner's Hope, and many others, are very encouraging prospects, with every indication of developing into paying properties.

MAYNARD DISTRICT.—This district is in the Hualapai mountains, twenty-eight miles east of Mineral Park. It is the finest wooded portion of Mohave, and is producing some very rich ore. The Atlantic and Pacific railroad will pass within ten miles of the mines. The American Flag is the leading mine of the district. It is a 2-foot vein of sulphuret ore, giving an average assay of $100 per ton. It is thoroughly opened by 2,000 feet of shafts and drifts. Some of the richest ore ever taken out in the Territory has come from this claim. It has produced $70,000, the ore being shipped to San Francisco. A mill will shortly be erected. The Antelope shows a 4-foot vein of fine sulphuret ore. It is opened by 400 feet of shafts and drifts. The mine has produced $15,000. The Dean has a large vein, nearly 6 feet in width. It has a shaft 180 feet, and 600 feet of tunnels. The ore is a sulphuret and of a high grade. The Mariposa is opened by 700 feet of shafts and drifts. It carries good ore and has yielded nearly $8,000.

CEDAR VALLEY DISTRICT is about sixty miles east of the Colorado river at Aubrey Landing, and about sixty miles south of Mineral Park. Wood is abundant, and water for ore reduction can be had at the Sandy, fifteen miles distant. The veins are well defined, in walls of granite. The ore is a sulphuret of silver. The Arnold shows a vein 18 inches wide, that assays $100 per ton. It has a shaft 60 feet, and a tunnel 130 feet. It is owned by the Arnold Mining Company, and has produced $20,000, gold and silver. The Silver Queen has a shaft 130 feet, and over 200 feet of tunnels and cross-cuts. Its vein is 3 feet, assaying $60 per ton. A 5-stamp mill and roaster have been erected on the property by the Hampden Mining Company. The Hibernia is a strong vein, 4 feet wide, with an average of $60 per ton. It has a shaft 100 feet. The Hope is a large vein and has some very rich ore. It is estimated that it has yielded $20,000. The Bunker Hill is a 2-foot vein, and the Congress is a vein of the same size, both carrying good ore. These are only a few of the mines of Cedar Valley. There are scores of others, well worthy of inspection.

HACKBERRY DISTRICT.—This camp is about 30 miles east of Mineral Park, in the Peacock range. The formation is a granite and porphyry. The camp was at one time the most prosperous in Mohave, but the stoppage of the Hackberry mine has caused it to become almost deserted. It is expected that with the advance of the Atlantic and Pacific railroad, which will pass within three miles of the mines, operations will again be resumed. The Hackberry vein is about 40 feet in width. About 18 inches of this vein carries rich silver ore, which gives an average, by working process, of $200 per ton. There is one shaft of 400 feet, another of 270, and one of 180 feet. The mine is opened by levels, drifts, and cross-cuts. It is estimated that the total yield of bullion has been over $300,000. A fine 10-stamp mill and roaster have been erected on the property. The mine is owned by the Hackberry Milling and Mining Company. The Descent is a small vein of rich ore, which has produced nearly $30,000. It has two shafts, one of 90, and one of 100 feet. The Hester is an extension of the Hackberry. It has two shafts, 100 and 60 feet each. It has produced about $10,000. The Hackberry South is a 4-foot vein, assaying $50 per ton. It is opened by several shafts, and has yielded $15,000.

SAN FRANCISCO DISTRICT is situated nine miles east of Hardyville on the Colorado river, in the Union Pass range. It was discovered in 1863, and work has been carried on there at intervals ever since. The Moss is the leading mine of the district. It is an immense gold ledge, nearly 40 feet in width, and will average $12 per ton, from wall to wall. The mine has been worked extensively in years past, and has produced some of the richest gold rock ever taken out in the Territory. It has one tunnel 290 feet, one shaft 240 feet, one shaft 98, and 1,700 feet of levels, drifts, etc. The mine has produced nearly $130,000. Its proximity to the river makes this a valuable property for those who have the requisite capital to work it properly. The San Francisco Moss is an extension of the Moss. It is a vein 40 feet in width, carrying ore that averages all the way across, $6 per ton. There are many portions of the ledge that go much higher. It has 300 feet of shafts, drifts, and tunnels. The West Extension is an 18-foot ledge of gold quartz, with a 60-foot shaft.

GOLD BASIN DISTRICT has just been organized, and is situated thirty-five miles north from Mineral Park, in the Cerbat range. The ledges are large gold-bearing quartz dikes. The El Dorado has a vein from 2 to 4 feet wide, that assays $40 per ton. The Northern Belle shows 2 feet that will assay $25 per ton. The Golden Rule is a vein about one foot wide, assaying $70 per ton. The Poorman has a foot of ore worth $60 per ton. The Indian Boy, Harmonica, O K, Antelope, Buckskin, and Banker, are all very fine-looking prospects, assaying from $15 to $100 per ton, in gold.

OWENS DISTRICT is in the southern portion of Mohave, near the line of Yuma. The formation of the country rock is granite and porphyry. Abundance of water is found in the Sandy, which flows through the district. The camp was established in the fall of 1874, and has been the most productive portion of

Mohave county. The fame of the McCracken has extended
all over the coast. The heavy cost of supplies of all kinds has
caused the temporary stoppage of work on this property, but
with the advent of the Atlantic and Pacific railroad, which will
pass within forty miles of the mines, work will no doubt be
resumed. The Alta and the Senator are the leading mines of
the district. They are on the great McCracken lode, which cuts
across the country for miles. They show veins of free-milling
ore, from 6 to 37 feet in width, which have worked $35 per
ton. Over $200,000 has been expended in work and in improve-
ments, and more than $800,000 in silver has been taken out.
They are owned by the McCracken Consolidated Mining Company. Two mills have been erected, one of 20 stamps and one of
10 stamps. The ores of this great fissure are mainly chlorides,
bromides, sulphides of silver, with some galena. Over 24,000
tons of ore have been extracted and worked. The mines are
opened by a shaft 367 feet deep, sunk on the line between them,
and by five adit levels run on the vein. In size of vein and
free character of its ores, the McCracken has few equals in the
Territory. The San Francisco and the Atlanta are north of the
Senator and Alta, on the same vein. They are owned by the
Peabody Mining Company, and have produced nearly $200,000.
They show about 30 feet of chloride ore, with some galena that
has averaged $30 per ton. The San Francisco has a shaft
300 feet, and over 300 feet of tunnels and drifts. The shaft of
the Atlanta is down 150 feet. It has also a tunnel 200 feet in
length. Work is carried on steadily.. The Centennial and the
Potts mine are about four miles south of the Senator, on the
same vein, have had considerable work done on them, and show
large ore bodies.

GREENWOOD DISTRICT adjoins Owens district on the east. Its
principal mine is the Burro, situated on a creek of the same
name. It is one of the largest veins in the Territory. It shows
35 feet of ore going from $8 to $300 per ton. A shaft has
been sunk 250 feet, and several cross-cuts made on the claim.
It carries gold and silver, and has abundance of wood and water
close at hand.

YUMA COUNTY.

The mineral field of Yuma county, in variety and extent, will
compare with any portion of the Territory. Gold, silver, copper, and lead abound in its mountain ranges. The history of
mining in this county dates back to 1858, when Colonel Snively
discovered the rich placers at Gila, twenty-five miles east of the
Colorado. For nearly four years work was prosecuted steadily
at this point, and a large amount of gold taken out. At Mesquite, some distance south of the railroad, very rich placer deposits have been discovered in the past year, and thousands of
dollars have been taken therefrom. In fact, that portion of
Yuma county south of the Southern Pacific railroad, is known
to be rich in alluvial gold, but, on account of the scarcity of
water, "dry washing" is the only way by which the mines can
be worked. The first mining north of the Gila river by Ameri-

cans began in 1862. In that year, Pauline Weaver discovered rich placers at a point seven miles east of La Paz. The fame of these discoveries spread far and wide, and within a year over 2,000 men were digging for the yellow treasure in the mountains east of the Colorado. It is estimated that gold to the value of over a million and a half of dollars was taken out. There are yet a number of Mexicans who stick by the old camp, and considerable gold finds its way to Yuma and other points. With the decline of the placer deposits, valuable discoveries of silver, copper, and lead were made in the mountain ranges that run parallel with the Colorado. Some of those discoveries have proven to be among the most valuable properties in the Territory.

CASTLE DOME DISTRICT is situated about twenty miles north of Yuma, in the Castle Dome mountains. The district was discovered in 1863, by the eminent geologist, Professor Blake, but owing to the hostility of the Indians, nothing was done until 1869. The mines are about seventeen miles from the river, and surrounding the lofty, natural "Dome," after which the range has been named. The formation is a slate and porphyry. The veins are found in fluor-spar and talc. The ores are a galena and carbonate of lead, carrying about $35 in silver, and from 60 to 70 per cent. in lead, with traces of gold. The ores are concentrated, hauled to the Colorado river, and shipped to San Francisco. The principal mines are the Railroad, Flora Temple, William Penn, Pocahontas, and Caledonia. They are owned by the Castle Dome Mining and Smelting Company, of New York. The Flora Temple has one main shaft 300 feet, and is thoroughly opened by drifts, tunnels, winzes, etc. The vein is about 4 feet wide, and the average yield is 30 ounces silver and 78 per cent. lead. The William Penn has two shafts of over 200 feet each, connected by a level 400 feet in length. It is a strong vein, showing good ore in every drift and stope. The yield is about the same as from the Flora Temple. The Pocahontas and Railroad have each a shaft 250 feet, and are connected by a drift 200 feet in length. These mines show large bodies of fine smelting ores, and go about 35 ounces, in silver.

The mines of Castle Dome are among the most productive and profitable of any in the Territory. Their proximity to the Colorado and the low rates of freight to San Francisco, permit the mining of ores of a low grade. The product finds a ready market in San Francisco on account of its fine smelting qualities, being used principally as a flux to more rebellious ores. It is estimated that these mines have already produced nearly $2,000,000 and from present appearances they promise to yield many millions more.

SILVER DISTRICT.—This district was first brought to notice nearly fifteen years ago by Colonel Snively, the discoverer of the Gila diggings. As placers were then the only mines which were thought worthy any attention, Snively and his companions abandoned the district and it remained undisturbed until about three years ago. At that time George Sills, Neil Johnson,

George W. Norton, and Gus Crawford relocated many abandoned claims and organized the district anew. Since then a great many discoveries have been made, some valuable properties have been developed, several important sales have been consummated, and Silver is to day the leading mining camp of the county. The district is situated on the Colorado river about forty miles above the town of Yuma, and about five miles from the stream. The formation of the country rock is mostly granite and porphyry, the surface showing some traces of volcanic action. The character of the ore may be generally described as an argentiferous galena, carrying rich sulphides, chlorides, and carbonates. The ore is generally found in combination with spar and quartz. There appear to be three main ore channels traversing the district, having a north-west and south-east direction. The veins are well defined and continuous, showing ore bodies of unusual width. The Red Cloud has the most development of any mine in the district. It is a large vein, 10 feet wide at bottom of shaft, with immense croppings. The mine was purchased from the original locators by the Red Cloud Mining Company, a New York incorporation. The company have sunk an incline following the dip of the vein 274 feet, and have started a working shaft which is down 300 feet. The vein is also opened by several open cuts and drifts. It is said that the yield of bullion has already reached $100,000. There is a 20-ton furnace at the river, five miles distant, which is working satisfactorily.

The Black Rock shows an immense outcrop, and appears on the surface to be nearly 200 feet wide. The property has been sold for $135,000. A shaft has been sunk 100 feet, following the foot wall, from which very rich ore has been taken. The purchasers of the property are pushing the work of development with energy, and the prospects for the opening of a valuable mine are not excelled anywhere. The Pacific adjoins the Black Rock, and is owned by the same company. It is a vein similar to the latter, showing fine ore. The Iron Cap has a shaft 200 feet deep, the vein between the walls in the bottom being fully 50 feet wide, showing good ore. The mine is owned by the Iron Cap Mining Company. The Silver Glance has produced some very rich mineral, and is one of the finest properties in the district. Like all the other veins, it is large and well defined. A tunnel over 200 feet in length has tapped the ledge nearly 150 feet below the surface. The Nellie Kenyon adjoins the Red Cloud on the north. The vein in some places shows a width of 30 feet. The ore is a rich galena, combined with fluorspar. By assay, the yield is 40 ounces in silver. The mine is comparatively unprospected, but gives every promise of becoming valuable.

About a mile to the eastward of the above mines is another ore channel, showing some fine-looking properties, among which are the Hamburg, Caledonia, Yuma Chief, and several others. The Caledonia has a shaft 100 feet, and carries a large vein of smelting ore. East of the last-mentioned group, about one

and a half miles, is the Klara camp, in which is located many
promising-looking claims. The Klara is a vein over 30 feet
wide, seamed throughout with ore. The property is being
thoroughly opened by shafts and cross-cuts. The Mamie
shows an ore body 15 feet wide, that gives an average assay of
40 ounces of silver per ton. The North Star is from 12 to 40
feet wide, and carries ore worth 30 ounces silver per ton. The
New York, Great Republic, Southern Cross, and many other
locations show large veins, although but little work has been
done upon them. Silver District has a desirable location. Its
veins are among the largest that have been discovered in Arizona. Its immense outcroppings show true fissures. Its ores
are easily reduced, and of a good grade. With all these advantages, there is no reason why it should not take a foremost
place among the bullion-producing camps of the Territory.

MONTEZUMA DISTRICT is five miles south of Castle Dome. The
veins are large, many of them being 40 feet wide. Assays as
high as 500 ounces, silver, have been made from several of them.
They carry gold also, and copper. Very little work has been
done in the district, but the surface prospects are most encouraging.

ELLSWORTH DISTRICT is about sixty-five miles from Sentinel
station on the Southern Pacific railroad, in the north-east corner
of Yuma county, and near the line of Yavapai. The mines are
situated in a rolling, hilly country, covered with a sparse
growth of grass. Mesquite, ironwood, and palo verde grow on
the hills, and water is found in sufficient quantities for the
milling of ores. The formation of the district is a granite and
porphyry. The veins are large, with bold outcroppings. The
ores of Ellsworth district are a gold quartz, carrying some silver.
The camp has a good situation, and will undoubtedly become
one of the leading gold camps of the Territory. The Oro
claim has a shaft 70 feet, besides open cuts and tunnels. It
shows 5 feet of quartz that has worked $20 per ton. The mine
is owned by the Oro Milling and Mining Company. A five-
stamp mill has been erected on the property and $10,000 has
already been taken out. The Nabob has a shaft 75 feet and a
body of quartz 4½ feet wide. Assays from this claim have gone
as high as $350 per ton. This is one of the most promising
mines in the district, showing large croppings and well-defined
walls. The Argenta has a vein 4½ feet wide, some of which
assays as high as $180 per ton. This claim carries a great deal
of galena, rich in free gold. The Socorro has a tunnel 50 feet
in length. It is a 4-foot vein carrying ore that goes $25 per
ton. The Richards and Ells claim is opened by a tunnel 100
feet in length. It shows 4 feet of ore, worth $20 per ton. The
Last Chance has a 20-foot shaft and shows an ore body 4½ feet
wide, that assays $29 per ton. The General Grant is down 20
feet, and has ore that goes $240 per ton. The Hawkeye, O K,
Peacock, Ellis, Oskoloosa, Oro Grande, Turtle, and many others,
all show good ore and large veins. But little work has been
done on any of them, but what has been done is sufficient to
prove their value.

PLOMOSA DISTRICT.—This district is about thirty-five miles east of Ehrenberg, on the Colorado river. It has been known since 1862, and has some large and rich bodies of copper and silver ores. A great deal of placer gold was taken from this neighborhood in early times. There is plenty of mesquite and palo verde growing on the hills, and water for milling purposes is only eight miles distant. The formation is granite, slate, limestone and porphyry. The Miami is an immense outcrop of gold quartz, running through a hill which is seamed with parallel veins its entire length. This ore body is about 300 feet wide. It has three shafts, 60, 50, and 40 feet, respectively. The ore is silver, carrying some copper. The Apache Chief is a vein 6 feet wide, assaying well in copper. A large amount of work has been done on the claim. It has a shaft 225 feet and a tunnel 100 feet, following the vein. The Pichaco shows 4 feet of galena ore that goes $50 per ton in silver. It has two tunnels, 100 feet each, and three shafts, the deepest being 100 feet. There are many other claims in this district well worthy of inspection by those looking for desirable investments.

HARCUVAR DISTRICT is situated about thirty miles north of Ellsworth, and about the same distance as the latter from the Colorado river. It contains several large copper veins, which show every indication of permanency. The country rock is granite. The veins average from 5 to 15 feet in width. Ores from this district have worked 37 per cent.

BILL WILLIAMS FORK DISTRICT.—This district is near the southern boundary of Mohave county, and extending west to the Colorado. The ores are copper and of a high grade. The Planet, the principal mine of the district, was discovered in 1863, and has been worked at intervals ever since, yielding over 6,000 tons of copper ore, going from 20 to 60 per cent. The ores from this mine have been shipped to San Francisco. The claim is opened by many shafts, drifts, and tunnels, and shows large bodies of ore. The Centennial and the Challenge copper mines, near the Planet, are also fine properties.

MARICOPA COUNTY.

Although generally considered an agricultural region, Maricopa county is rich in the precious metals, almost every mountain range within the limits of the county showing mineral. The north-eastern portion, embracing the spurs and foothills of the Superstition and Mazatzal ranges, is known to be rich in gold, silver, and copper, but as yet has been but little explored. That division of the county, south of the Gila, is known to contain rich silver and copper deposits, although the development thus far has been very slight. Maricopa possesses every natural auxiliary for the mining and reduction of ores, besides producing all the supplies necessary for the successful prosecution of the industry.

The Vulture mine is situated in the north-western portion of the county. This great lode has a reputation which has made it famous all over the Pacific coast. No mine ever located in the Territory is perhaps so well known beyond its borders. The mine was discovered in 1863, by Henry Wickenburg, and worked almost continuously by an Eastern company until 1873. The high rates of freight and the cost of hauling the ore—$8 per ton—to the mill, sixteen miles distant, caused a suspension of work and an abandonment of the property. The mine was afterwards located by other parties, who erected a ten-stamp mill on the Hassayampa, twelve miles distant, and worked the ores successfully for several years. Three years ago the property passed into the hands of the Central Arizona Mining Company, and since that time the mine has entered on an era of prosperity it never knew before. The new company have brought water in pipes from the Hassayampa, a distance of sixteen miles, and have erected an eighty-stamp mill at the mine. The property has had more work done upon it than any mine in the Territory. A deep excavation on the surface shows the ore body to be nearly 100 feet in width. A depth of 390 feet has been reached, and several levels and cross-cuts run on the vein. The ledge lies between a hanging wall of porphyry and a foot wall of talcose slate. It is situated in a low hill, and at a depth of about 200 feet the vein is almost vertical. With the present arrangements for reduction, the ore is extracted and milled at a total cost of $2 25 per ton. More stamps will soon be added, and the yield of bullion largely increased. The Vulture has produced more money than any mine in the Territory, the total yield being placed at $3,000,000 in gold. With the immense ore bodies in sight, and the appliances for reducing them, we may look to see many millions more taken from this fine property.

CAVE CREEK.—This district is about thirty miles north from Phœnix, in the southern spurs of the Verde mountains. The country rock is slate and granite; the veins are of good size, with well-defined walls. Water is found in abundance from three to five miles of the camp. The Carbonate Chief shows a vein of carbonate ore nearly 7 feet wide, assaying $50 per ton. It carries gold and silver, and is opened by a shaft 50 feet in depth. The Panther is a large vein, with ore similar to the Chief. It is opened by several shafts and tunnels. Both of these mines are owned by the Panther Mining Company. The Lion is a 4-foot vein of gold quartz. Ore from this mine has worked $40 per ton. It is opened by a shaft 30 feet and a drift 50 feet, and has produced $10,000. The Rackensack is a 2-foot vein, going $40 per ton in gold. It is opened by a shaft and tunnel, the former 50 feet, and the latter 60 feet. It has yielded $8,000. The Golden Star is a fine-looking body of quartz. It has a shaft 60 feet deep, and has produced about $10,000. A ten-stamp mill has been erected on the claim. The Hunter's Rest, Maricopa, Chico, and Catherine are all promising prospects, showing large ore bodies.

WINNIFRED DISTRICT.—This district is about fifteen miles north of Phœnix. The ledges are a gold quartz. The country formation is a granite and slate. A five-stamp mill run by water-power, has been erected on the Grand canal, four miles from Phœnix and eleven miles from the mines, where the ore is reduced. This is a new camp, but promises to become an important one. The Union is opened by a 75-foot shaft, and shows a vein 3½ feet wide, all of which has worked $15 per ton. This mine is worked steadily, and promises to become a valuable property. The Scarlet has a vein 3 feet wide, assaying $50 per ton. It has a shaft 20 feet deep. The Gila Monster shows 2½ feet of good ore. The claim is opened by a 40-foot shaft. The Red Dog, San Diego, and Mogul are all fine prospects.

MYERS DISTRICT.—This district is about forty miles south of the Gila Bend station on the Southern Pacific railroad. The ledges show strong and well-defined fissures filled with argentiferous galena and carbonate ores, assaying all the way from $50 to $5,000 per ton. Wood and water are not plentiful. The principal mines are the Gunsight, Silver Girt, Morning Star, Crescent, Monumental, and Atlanta. Some rich copper discoveries have been recently made in the mountain range south of Phœnix. The ledges are represented as being from 10 to 30 feet wide, carrying ore which assays from 20 to 53 per cent. But little work has yet been done on these veins, but they give every promise of becoming productive copper properties.

GRAHAM AND APACHE COUNTIES.

Although these divisions of the Territory have not heretofore received that attention from mining men which the richness and extent of the mineral fields have deserved, it is well known that gold, silver, copper, lead, iron, coal, and other minerals exist throughout their mountain ranges. Their remoteness from the traveled highways, and the difficulties and cost of procuring supplies and material, are the causes which have retarded the development of the mining interests of these counties. The streams throughout the Sierra Blanco range contain placer gold in large quantities, and have a sufficient supply of water to make mining for the metal, with proper hydraulic machinery, profitable. Although the formation in this portion of the Territory is of an eruptive character, there are stretches of the primitive rock in many places, giving every indication of containing mineral. But little prospecting has been done in Apache county; but the building of the Atlantic and Pacific railroad through its center will no doubt give an impetus to this as to all other branches of industry. With its great coal-fields and salt deposits, of which we shall speak hereafter, no county in the Territory has greater natural facilities for ore reduction, and no portion of Arizona presents a more inviting field for the searcher after the hidden treasures.

Graham is the youngest born of the counties of Arizona, and promises to become one of the richest in its mineral possessions. It can show the most productive copper mines in the Territory, if not in the United States. Gold, in alluvial deposits and in quartz ledges, is found in many of its mountain ranges, while silver and coal are likewise among its resources. There is a large portion of Graham, not yet prospected, which gives every indication of being mineral-bearing.

The famous Longfellow copper mines are in Graham county. They are situated on the San Francisco river, a few miles above its junction with the Gila. This region was known to be rich· in copper, but it was not until 1874 that mining was carried on to any extent. Before the building of the Southern Pacific railroad, the copper mat was shipped a distance of 700 miles by wagons to the nearest railroad, and from there forwarded to Baltimore. Notwithstanding the enormous cost of this mode of transportation, the ore paid its owners a profit. The deposit appears to be a regular mountain of ore, drifts and tunnels having, so far, failed to find anything like a wall; and in whatever direction the workmen penetrated, they have encountered the ore body. As a consequence, the mine resembles in some respects a quarry, showing metal in every direction. The property is owned by an incorporated company, which appears to be a very close corporation, not disposed to let outsiders know too much about the "good thing" they possess. The ore is copper glance, red oxide, and a carbonate. Extensive reduction works have been erected on the San Francisco river. The yield is about 14,000 pounds daily, which will soon be largely increased by additional reduction facilities. What the total yield from these mines has been has not been ascertained, though it is known to reach up into thousands of tons. The company give employment to a large number of men, and a flourishing camp, known as Clifton, has sprung up near the mines. The Detroit Mining Company, operating three miles from the Longfellow, have opened up a splendid property. They are putting up reduction works, and intend to connect their mines by a branch road to the Southern Pacific. The ore is equally as rich as the Longfellow, and quite as extensive, and will no doubt prove as productive. There are many other copper properties in this region, which give every promise of becoming valuable.

The rich gravel deposits of the San Francisco river are the most extensive in the Territory. A Boston company have recently purchased nearly 1000 acres of this gravel bed, and are making preparations to work it on a large scale. Fifteen miles of piping have already been laid, and hydraulic machinery will be erected at once. These gravel beds have been thoroughly prospected by shafts and tunnels, and show gold in paying quantities in every foot. In the eastern portion of Graham, and lapping over into Pinal, is De Frees district. It is about ten miles south of the Gila, in the Pinaleno mountains, and about sixty miles from the Southern Pacific railroad. The camp has plenty of wood, water, and fine pasturage. The ledges carry silver and copper. The formation is a lime and porphyry. But little work has yet been done, but the showing is most encouraging. The principal mines are the Fairy Queen, a 4-foot vein of carbonates, assaying $40 per ton; the Nez Perces, a 6-foot vein, giving assays of $60 per ton, and opened by a forty-foot shaft; the Silver Glance, showing 2 feet of ore that assays $80 per ton. The Charter Oak has a shaft 35 feet, and a 4-foot vein giving $50 per ton. The Calypso, Ulysses, Ironclad, Shotgun, and Iron Cap, are all encouraging prospects.

COAL AND SALT.

Besides its gold, silver, copper, and lead, Arizona possesses immense coal-fields and large salt deposits. The latter article is an important factor in the reduction of silver ores, and a prime necessity for their successful treatment. Arizona, in this respect, is endowed beyond her neighbors, and nature, while scattering in profusion her mineral wealth throughout the Territory, has also provided the agents for its successful working. About 100 miles above Phœnix, on Salt river, there is a high bluff composed almost entirely of salt. From this bluff, several springs highly impregnated with saline matter, flow into the stream. The river above this point is pure and clear, but below it has a strong brackish taste. The salt is of a fine quality, being remarkably free from soda, gypsum, and other impurities. An effort to erect a factory, and bring the article into market, has not proven a success, owing to the expense and difficulty of getting material on the ground. . This deposit is an extensive and valuable one, and will yet prove a lucrative investment for those who have the requisite capital. Near Camp Verde, in Yavapai county, there are several large salt bluffs or hills. This salt carries large quantities of soda and magnesia. It is used by cattle raisers for salting their stock; the supply is almost inexhaustible, and the salt could easily be freed from its impurities, and made to answer all purposes, dairying, table use, or the working of ores. Salt lagoons are met with in several places in Apache county. The principal lake or lagoon is near the line of New Mexico. About 1,000,000 pounds are taken yearly from this lake, and with proper facilities it could be made to produce an almost unlimited supply. The salt is precipitated to the bottom of the lake, wagons are driven into the shallow water, and the glittering crystals shoveled in. This is one of the most valuable salt springs on the continent, and besides supplying cattle raisers in Apache and portions of Yavapai, furnishes large quantities for the working of silver ores. The Atlantic and Pacific railroad, passing within a short distance north, will be the means of providing a larger market for this valuable article.

Next to the great fields of Pennsylvania, there is no portion of the Union which can show such immense coal measures as Arizona. This coal region embraces the northern division of Apache, and that portion of Yavapai north of the Little Colorado. The coal-field extends into New Mexico on the east, and Utah on the north; competent geologists have estimated its area at over 30,000 square miles, or more than half the coal measures of the United States. The beds vary in size, from two inches to thirty feet. A gentleman who visited these coal-fields in 1873, writes of them as follows: "Close to Fort Defiance a vein exists nine feet thick, and it seems to possess all the qualities of excellent bituminous coal, and to rank next to anthracite for domestic purposes. * * * I see no reason why it should not be pre-eminently useful for generating steam and for smelting ores. * * * This description will apply to all the coal in the great Arizona coal basin. * * * The next great bed of coal encountered is situated about twenty miles north-west from the Moquis villages, and close to the northern verge of the Painted Desert. * * * It is twenty-three feet thick and boldly crops out for a distance of three miles. This coal is close, compact, and close burning; melts and swells in the fire, and runs together, forming a very hot fire, and leaves little residuum. It resembles, in external appearance, the Pennsylvania bituminous coal. * * * The trend of the coal-beds is north and south, and overlying this great deposit is drab clay, passing up into areno-calcareous grits, composed of an aggregation of oyster shells, with numerous other fossils which must have existed in this great brackish inland sea about the dawn of the tertiary period, probably in the eocene age."

A peculiarity of this great coal region is the number of petrified trees which are found all over its surface. Whole forests of these petrifactions are met with in all directions, proving that in ages past the country was covered with a dense growth of timber. Some of these trees are three feet in diameter and from fifty to sixty feet in length. The railroad on the thirty-fifth parallel will pass south of this immense coal deposit, and a branch will no doubt tap it. There is here coal enough to supply the United States for ages to come.

Bituminous coal of an excellent quality has recently been discovered on Deer creek, a tributary of the Gila, and near to the point where it enters that stream. The mines are in Pinal county, and about twenty miles east of the mouth of the San Pedro. The field, as far as has been ascertained, is about three miles long and two miles wide. The veins are from three to eight feet thick; the coal makes an excellent coke, and for domestic purposes it is said to be unequaled. The coal-beds are about sixty miles north of the Southern Pacific railroad, and arrangements are now being perfected for the running of a branch which will open a market for this valuable deposit. Coal of a fine quality has been found near Camp Apache and at other points in the eastern part of the Territory, but no effort has yet been made in the way of development.

It will thus be seen, from this hasty glance at the coal-fields of Arizona, that there is here abundance of the article, and of a good quality. If anything was wanting to make this Territory the greatest mining region on the globe, these vast coal deposits supply that want, and contain an inexhaustible fuel supply for the working of its ores, and for all other purposes to which it may be applied.

BULLION YIELD.

No truer test of the richness of Arizona mines can be found than in the steadily increasing volume of bullion which is finding its way out of the country. This yield has more than doubled each year since the Southern Pacific railroad entered the mineral fields of Southern Arizona. In 1880, according to the report of Wells, Fargo & Co., the total output was near $4,000,000. This did not include the raw ores, concentrations, and placer gold shipped through other sources. The yield for 1881, reckoning on the basis of the present monthly production, will be about $9,000,000. This estimate does not include the copper product, which will reach 4,000 tons, worth over $1,500,000. Add to this the ores and concentrations and placer gold, which finds its way out of the country, and the entire bullion yield for the present year will be over $12,000,000, thus placing Arizona third on the list of the bullion-producing States and Territories.

This is a good showing for a country whose total shipment six years ago amounted to only $109,083! With such a rapid increase in the output of treasure since the building of one railroad, what may we not look for a year or two hence, when Northern Arizona will be opened by another transcontinental line, and the leading camps north and south will be tapped by branches? It is not too much to expect that the Territory, now third on the list, will take the first place in the production of the precious metals. No country has so extensive a mineral field, possesses so many natural advantages, or can show ores of such wonderful richness, lying almost at the very surface. Some of the most eminent geologists and mineralogists long ago predicted that the region now embraced within the Territory would yet prove to be the richest mining country on the globe. The soundness of their judgment is at last being practically demonstrated, and it is evident that Arizona is soon to become the great bullion-producer of the world. No mining country can show such a return for the amount of capital invested; none presents to the man of enterprise more guaranties for success, and none has so bright a future.

AGRICULTURE AND GRAZING.

Although not generally considered an agricultural country, Arizona contains some of the richest valleys to be met with in the United States. Cereals, fruits, and vegetables of all kinds are raised in every portion of the Territory. Wherever water can be had for irrigation, a bounteous yield is assured, and in the southern portion of the Territory, two crops in the same year are not uncommon. The farming land of Arizona is confined at present to the valleys of the principal rivers. There are millions of acres among the hills and *mesas*, with a fine soil and agreeable climate, capable of producing anything grown in the temperate or semi-tropical zones, with a sufficient water supply. It is believed that by the sinking of artesian wells, much of this land can be brought under cultivation, and what are now barren, dreary wastes, be changed into fields of waving grain, with comfortable homes, embowered in refreshing shade. Artesian water will confer untold benefits on the Territory, and no wiser or more beneficial measure could receive the support of the general government. There are about 45,000 acres under cultivation in the Territory at the present time, and there are still thousands of acres unoccupied in the valleys of the Colorado, the Gila, and the Salt rivers. The valley of the Colorado, containing the richest land in the Territory, is subject to annual overflows, and has been farmed only to a limited extent by whites. The valley of the Gila is settled from the line of New Mexico to its junction with the Colorado. It contains some of the best land in the Territory, and produces large crops of grain, fruit, and vegetables. There are large tracts of land in this valley at Gila Bend, and at other points where there is sufficient water, which are open to pre-emption. Salt River valley contains the finest body of agricultural land in the Territory, and produces two thirds of all the cereals grown in the country. There is plenty of water at all seasons, and the system of irrigation is more extensive than in any other portion of Arizona. As farming will always be a lucrative business in the Territory, owing to the limited area of land which can be brought under cultivation and the large population who will be engaged in mining, an impartial statement of its capabilities as an agricultural country are here given for the benefit of those who are thinking of coming hither and engaging in this business; and for the purpose of setting forth this information more clearly, the agricultural resources of the different counties are given separately.

MARICOPA COUNTY.

This county has been well· named "the garden spot of the Territory." It has the finest body of land in Arizona, and its farms, orchards, and vineyards will not suffer by comparison with any portion of the Golden State. The first settlement was made in this valley a little over ten years ago. It was then a barren desert, covered with coarse grass, sage, and cactus; to-day it is one of the loveliest spots on the Pacific coast. Fields of golden grain and blossoming alfalfa; extensive vineyards and orchards; beautiful gardens, brilliant with their floral adornments nearly every month in the year; groves of cottonwoods and lines of the graceful Lombardy poplar diversify the landscape in every direction; and to crown all, tasteful homes are seen peeping above their leafy surroundings up and down the river as far as the eye can reach. In this beautiful and productive spot, wheat, barley, and alfalfa are the principal crops. The soil is a sandy loam, but there are portions of the valley which are a heavy, rich adobe. Up the river, near Mesa City, the soil is light, but well adapted to fruits. Besides its large crops of grain, Maricopa produces the finest vegetables in the Territory. Pumpkins, squashes, onions, turnips, cabbages, watermelons, and everything in the vegetable line, are raised in large quantities, and are in market by the first of March.

The soil is peculiarly adapted to the raising of sugarcane, and some of the stalks attain a height of over twelve feet. It has been estimated that an acre of this cane will yield 200 gallons of syrup, of an excellent quality; it also makes a nutritious food for horses and stock. There are about 1,000 acres of this valuable plant now under cultivation, and the area is being steadily increased, many farmers finding it more profitable than the raising of grain. Figs, peaches, apricots, and grapes do well in the Salt River valley, and in size and flavor are not excelled on the Pacific coast. Apples and strawberries are cultivated to some extent, and experiments with oranges, lemons, and other semi-tropical fruits, have shown that the valley is peculiarly adapted for their successful cultivation. In fact, there is no country west of the Rocky mountains which seems so well fitted for the raising of fruits. Climate, soil, and situation, all seem to be favorable, and the valley promises to become one of the greatest fruit-raising regions of the Pacific coast. The business of wine-making is being gone into extensively, and a very fine article is produced, which in body and flavor compares favorably with the best California. There are at present 500 acres in grapes, 150 acres in peaches, 50 acres in apricots, 25 acres in figs, besides a number of acres in apples, strawberries, oranges, lemons, etc. Of barley, it is estimated there are over 5,000 acres in cultivation; in wheat, 5,000 acres; corn, 500 acres; and alfalfa, 2,000 acres. The average yield of wheat and barley is about 1,500 pounds to the acre, and the average price received by the farmers is about $1 40 per hundred, sacked.

The grain is sown in the Salt-river valley in October, November, and December. Harvesting begins in the latter part of May, and ends the first of July. Everything is grown by irrigation. From three to five floodings are necessary to raise a crop of small grain. The cost of irrigation is about $2 50 per acre. Where the land is favorably situated, it is estimated that crops can be raised as cheaply by this plan as by rainfall, besides being much more certain. The water is conveyed over the land by large canals. Owing to the number of these canals, a large quantity of water is wasted and lost by evaporation, which could be utilized with a proper and comprehensive system of irrigation. The farms in the valley extend for nearly 30 miles along the river. The amount of land which can be cultivated depends entirely on the supply of water. There are at present something over 16,000 acres reclaimed from the desert; with a proper irrigating system, it is believed that as many more can be made productive. Land in the valley, with a water right, can be bought for $5 and $10 per acre, according to quality and situation.

Maricopa county, besides the valley of the Salt river, has some fine farming land along the Gila, which is cultivated at several points. The land is fully as rich as that on the Salt, but the supply of water is not as abundant. At Gila Bend, below the junction of both streams, there is a fine body of land, capable of producing all kinds of grain, fruits, and vegetables. Most of this land is still unoccupied, and open for pre-emption. Maricopa will always be the leading agricultural county of the Territory, and in a few years the region of country which has Phœnix for its center will become one of the most inviting and productive spots on the coast, rich in its immense fields of grain, and beautiful with its groves of orange trees, and its vineyards and orchards.

PINAL COUNTY.

The agricultural land in this county is confined to the valleys of the Gila and the San Pedro. For a distance of eighteen miles along the former stream there is a line of fine farms, and for thirty miles up the San Pedro, the valley has been brought under cultivation at different points. In the neighborhood of Florence, the county seat, the valley of the Gila is over a mile wide, and contains some of the richest land in the Territory. Here, as everywhere else, irrigation is required to produce a crop, and the area that can be cultivated depends entirely on the water supply. Corn, wheat, barley, alfalfa, vegetables, and fruits are raised in Pinal county. The soil is a rich loam of durable fertility, and well adapted to the usual agricultural products and semi-tropical fruits. There is no more beautiful sight in the Territory than the valley of the Gila surrounding Florence, when the ripening grain, waving fields of alfalfa, and shady groves of mesquite and cottonwood are in their bloom. There are thousands of acres of fine land above and below Florence, which are lying idle for the want of water. It is believed that with a proper system of irrigation, double the number of acres now under cultivation could be made to produce fine crops. There is evidence in the ruins of the Casa Grande that this portion of Arizona supported a dense population at one time; and the remains of the large irrigating canals go to show that those ancient tillers of the soil had a much more comprehensive idea of the irrigating problem than their modern successors. The number of acres under cultivation in Pinal county is estimated at 6,000, not including the land occupied by the Pimas, which is nearly all within the limits of this county. The yield for 1880 was: Barley, 1,000,000 pounds; wheat, 400,000 pounds; corn, 350,000 pounds; besides large quantities of hay and alfalfa. The yield of grain to the acre was: Barley, 1,500 pounds; wheat, 1,200 pounds; besides cereals, beans, potatoes, onions, cabbages, turnips, and all kinds of vegetables are raised in abundance.

Peaches, grapes, apricots, pears, figs, quinces, and pomegranates, all do well in Pinal, and many farmers are going into the business extensively. The climate and soil are specially adapted for fruit culture, and the valley of the Gila yet promises to become one immense orchard and vineyard.

YAVAPAI COUNTY.

The principal body of farming land in this county is found along the valley of the Verde. This valley averages from a few hundred yards to a half a mile in width. The soil is a rich loam, and in places a black mold of great fertility. The river bottom is settled its entire length, where it is not confined to canyons. There is plenty of water for irrigation, and good crops are raised in the driest season. Corn, wheat, and barley are the principal productions. Although but little attention has been paid to fruit, it has been demonstrated that fine grapes and peaches can be grown in this valley. Outside of the Verde the farming lands of Yavapai are confined to small valleys situated from four to six thousand feet above sea level. Among the most important of these valleys are Williamson, Chino, Peeple's, Agua Fria, Skull, Kirkland, and Walnut Grove. Their soil is generally a rich mold, formed by the detritus from the surrounding hills. There is no water for irrigation in most of them, and farmers depend entirely on rain for the raising of a crop. Corn, wheat, barley, alfalfa, and all kinds of vegetables, are raised in these elevated valleys, their greatest drawback being late and early frosts and droughts. Fine apples and peaches are grown in several places, and grapes in some secluded nooks. The number of acres under cultivation in Yavapai is estimated at 5,000, although no reliable data can be had from the assessor's office.

PIMA COUNTY.

The valley of the Santa Cruz is the principal agricultural settlement of this county. This stream, which rises in the Huachuca mountains, sinks in the thirsty sands for more than two thirds of its course. Near Tubac and Calabasas, opposite Tucson, and at San Xavier, the stream comes to the surface, and the land in the vicinity is brought under cultivation, producing crops of cereals, vegetables, and fruits. The valley of the Santa Cruz, opposite Tucson, has been cultivated for hundreds of years, and shows no diminution in its productiveness. The soil is rich, and only needs water to grow anything that is planted in it. The Sonoita valley, east of the Santa Ritas, and about sixty miles south-east of Tucson, is one of the most productive spots in the southern portion of the Territory. It extends from old Fort Buchanan to Calabasas, nearly thirty miles, and is settled, wherever water can be had, the entire distance. The soil is a rich, dark loam, and the climate is well adapted for fruit raising. This valley was time and again swept with fire and drenched with blood during the Apache wars, and the graves of its early settlers mark the hillsides from one end of the valley to the other. The valley of the Arivaca, in the southern part of the county, contains some good land, but it is claimed by a "grant," thus preventing settlement.

The yield of cereals in Pima county for the year 1880, was as follows: Wheat, 1,000,000 pounds; corn, 500,000 pounds; barley, 1,000,000 pounds. This yield includes the products of the farming lands now embraced within the boundaries of Cachise.

CACHISE COUNTY.

The agricultural resources of this county are confined to the valleys of the San Pedro and the Babocomari. The former stream rises in Sonora and flows through Cachise and Pinal counties into the Gila. The valley of the San Pedro, in its upper course, is sometimes a mile in width, and the soil is of an excellent quality, capable of raising all kinds of grain and vegetables. That portion of the valley near the line of Sonora is claimed by a "grant," and is devoted entirely to grazing. No figures have been received as to the number of acres under cultivation and the grain yield of this county.

GRAHAM COUNTY.

This county, which embraces the upper valley of the Gila, contains a large body of fine farming land, with plenty of water for irrigation. The Pueblo Viejo valley, which supported a dense population in times past, is yet rich and productive, yielding large crops of corn, wheat, barley, alfalfa, and vegetables. It is estimated there are 10,000 acres under cultivation in Graham. Large tracts, now lying idle, can be made productive by extending the present irrigating canals. The soil of this portion of the Gila valley is similar to that near Florence. Fruits of all kinds do well in this region, and no finer potatoes are raised in the Territory. The first settlements were made in this valley in 1872, and at the present time it is, next to Salt river, the largest producer of cereals in Arizona. There is here an opportunity to secure a comfortable home in a fine climate, and near to a profitable market.

APACHE COUNTY.

This county has some good land along the Little Colorado and its upper tributaries. From Springerville to Brigham City, the valley has been brought under cultivation wherever water can be obtained. Several Mormon colonies have settled in this region, and have raised good crops of corn, wheat, and barley, besides fine vegetables. This part of Arizona is prolific in its growth of wild flax. This fact arrested the attention of the Spanish explorers, who called the stream Rio de Lena, or Flax river. No effort has been made to cultivate this fiber, but it is believed it will yet become an important branch of industry.

GILA COUNTY.

Very little farming is done in this county. With the exception of a few gardens along Pinal creek, and a narrow strip on Salt river, there is no land within its limits—if we except the San Carlos Indian reservation—which has sufficient water to produce crops. There is some fine valley land, with rich soil and a delightful climate, which could be made to yield bountifully by the aid of artesian water.

YUMA AND MOHAVE COUNTIES.

These two counties embrace the great Colorado valley, which contains thousands of acres of the richest soil in the United States. Owing to its yearly overflow, the valley is covered with a coating of vegetable mold, which constantly enriches the soil. Vegetation is very rapid in this valley. Weeds, grasses, and wild hemp attain an amazing height in a few weeks after the waters have receded. In fact, everything grows in tropical luxuriance. If kept from overflow, no better soil for cotton, sugar, hemp, and semi-tropical fruits is found on the continent. In some places the bluffs come down to the stream, and at other points the valley is from one to five miles wide. Below Ehrenberg, the area of valley land is much greater than above. To bring the waters of the Colorado by canals over its rich valley and prevent the river from overflowing, would no doubt be an expensive undertaking, but the hundreds of thousands of acres of magnificent land which would thus be reclaimed are a prize worth striving to gain. A company has been formed for the raising of hemp and sugarcane, which has already begun operations in the valley below the town of Yuma; but with the exception of small patches cultivated by the Indians, the rich valley of the Colorado is still virgin soil. Between the junction of the Gila and the Colorado, there is a tract of very rich bottom, by some estimated at 30,000 acres, all of which could be brought under cultivation at a moderate cost. The two largest streams of the territory, flowing on either side, would give an inexhaustible water supply, and the configuration of the ground is such that it can be easily irrigated. There are several fine ranches along the valley of the Gila, in Yuma county, which yield good crops of grain and vegetables. The total number of acres under cultivation in the county is about 2,500. The valley of the Colorado, in Mohave county, presents the same features as in Yuma, but is not so extensive. The soil is equally as rich and productive, but it requires capital to open canals, throw up embankments, and put the land in a condition for successful cultivation. At present farming in Mohave is confined to the Big Sandy, in the southern part of the county, where there are about 1,000 acres under cultivation, producing fine crops of grain, vegetables and fruit.

From this brief summary, it will be seen that successful farming in Arizona depends entirely on irrigation. No finer crops are raised in any country than in this Territory, where water can be had. There are thousands of acres of productive land in the leading valleys, which can be made available by a proper distribution of the present water supply. While the wealth of Arizona is in its mines, agriculture will always be a profitable calling, and the products of the soil command a good price. There is no land more prolific, no climate more equable, and no country where the labors of the husbandman receive a more generous reward. In saying this, however, it is not the intention to invite hither a large agricultural population. As has been before stated, the area of land which can be brought under cultivation is limited, and must remain so until artesian water shall send forth its fructifying streams, and make the dry valleys and plains to blossom as the rose.

GRAZING.

Within the last few years, cattle raising has become an important industry in Arizona. The fine grasses and the delightful climate make this region the very paradise of the stock grower. All the year round the rich grasses cover mountain, valley, and *mesa*. Situated between the extremes of temperature, subject neither to the fierce "northers" of the South-west, nor the heavy snows of more northern latitudes; requiring no expensive outlay for the protection of stock in winter, and with a range which is only limited by the boundaries of the Territory, there is no portion of the United States which presents a finer field for the successful prosecution of this industry than the Territory of Arizona. Prior to the year 1874, the business was attended with many difficulties and dangers. The marauding Apache was always ready to swoop down on the flocks and herds of the settler, and the industry was confined to the immediate vicinity of towns and military posts. Since the "disturbing element" has been placed on reservations, stock growing has made rapid strides, and large bands of cattle and sheep are found in all portions of the Territory. No finer beef is raised in the United States than is produced in Arizona. The rich gramma grass which covers its valleys and hills, is unexcelled for its fattening qualities, and the sweetness and flavor which it imparts.

As with agriculture, the sinking of artesian wells-will be of great benefit to the stock interests. There are millions of acres of fine grazing land now lying idle which could be made to sustain thousands of cattle if water could be had. That flowing water can be found in these valleys is almost certain. Surrounded as they are by lofty mountains, and forming natural reservoirs for the moisture which falls upon them, they offer every encouragement for the sinking of wells, and give almost certain guaranties of producing an abundant supply. Hitherto no effort has been made in this direction, owing mainly to the fact that the grazing lands adjacent to the streams and living springs have furnished an abundant supply for the stock already in the Territory. Besides the home market, which is steadily increasing, the building of two transcontinental railways opens to the stockmen of the Territory the marts of the Atlantic and the Pacific, and of Europe. Beef is shipped from the northern Territories to England, at a good profit, and there is no reason why Arizona should not be able in a short time to supply the epicures of the British Isles with a sample of beef far superior to the stall-fed article on which John Bull has so long prided himself. In fact, there is no branch of industry in the Territory which offers superior inducements for investment than the cattle business, nor is there any State or Territory in the Union where this business can be carried on with less expense or liability to loss. While thousands of cattle and sheep are annually destroyed by cold and snows in northern latitudes, cattle graze on the mountains, hills, and valleys of this favored land every month in the year.

What has been said of cattle will also apply to sheep. The mutton from Arizona grasses is noted for its fine flavor and tenderness. The wool is of a prime quality, commanding the highest price paid for the Pacific coast product. A superior breed has been introduced within the past three years, and the yield has correspondingly increased. The high rates of freight have been the great drawbacks to successful sheep raising, but the railroads have done away with all this, and the sheep industry of Arizona is one of the most lucrative branches of business in the Territory. All over Northern Arizona the short sweet grasses that grow on the *mesas* and mountain sides make an excellent feed for the animal, and in many of the valleys, the *alfileria*, or wild clover, has been introduced by sheep driven from California, and is attaining a strong and thrifty growth. Sheep are sheared twice a year, the average yield per head being about six pounds. The grazing grounds of Yavapai county are among the richest in the Territory. The snowfall of winter and the rains of summer cover the whole region with a heavy growth of fine, nutritious grasses, which keep stock in prime condition. The whole of the Great Colorado plateau, in Yavapai and Apache counties, affords one of the very best stock ranges to be found in the western country. The great table lands and spurs of the San Francisco, Sierra Blanco, and Mogollon ranges, are at all times covered with a heavy growth of gramma and other grasses, while the climate is especially salubrious, being removed from the oppressive heats of summer and the heavy snow storms of winter.

Pima county has large tracts of excellent grazing land along the Santa Cruz, and in the rolling, grassy country south and east of Tucson. Large herds of cattle cover these plains and hillsides, and keep in prime condition at all seasons. This county is also an excellent sheep range, and ships large quantities of wool. The building of the Southern Pacific railroad has opened new markets for the beef and wool of Southern Arizona, and the number of stock is increasing rapidly. Cachise county, formerly a part of Pima, has extensive ranges in the San Simon, Sulphur Spring, and San Pedro valleys. Although no data have been received from this county, it is known that it contains a large number of sheep and horned cattle.

The central counties, including Graham, Pinal, Maricopa, and Gila, embrace some magnificent grazing lands, thousands of acres of which are unoccupied. The Arivaypa valley, in

Graham county, supports large bands of horses and cattle; the grasses in this region are not excelled in the Territory, and the quality of beef produced has no equal in the western country. To describe fully in detail all the ranges in the Territory would require much more space than can be given in a compilation of this nature. Speaking in general terms, it can be truly said that there is no better grazing region west of the Rocky mountains than Arizona; and while the want of water prevents many portions of the country from being occupied, there is yet room for thousands of cattle and sheep where water is abundant, where animals keep fat winter and summer, where the climate is all that could be desired, where disease is unknown, and where an energetic man with a small capital, who understands the business, can make himself independent in a few years.

Below is a statement of the number of cattle and sheep in the several counties at the present time. No figures have been received from Gila or Cachise, and consequently we are unable to give the number in these counties:

Cattle.

Yavapai	27,528
Pima	18,000
Graham	12,500
Maricopa	6,000
Pinal	5,000
Apache	10,000
Yuma	4,000
Mohave	5,500

Sheep.

Apache	300,000
Pima	50,000
Yavapai	28,316
Maricopa	15,000
Pinal	2,000
Graham	13,000

Besides cattle and sheep, Graham county has 2,500 head of horses and mules; Pima has 2,000 head of horses and 500 head of mules; Maricopa has 1,000 head of horses, about 500 head of mules, and over 4,000 hogs. The extensive alfalfa fields of this county afford excellent feed for hogs, the business is being gone into on a large scale, and home-made bacon, equal to the best California, is put up in the Salt-river valley. Yuma county has about 1,500 head of horses and mules; Mohave county has 600 head of horses, and 150 head of mules; Yavapai county has 3,815 head of horses, 627 head of mules, and 1,500 head of goats; many of the latter are pure-blood Angoras, and appear to do excellently well in this climate.

The live stock throughout the Territory is being steadily increased and improved by the importation of pure breeds, and in a few years we may expect to see the immense stretches of grass lands, now unoccupied, covered with thousands of cattle. horses. and sheep.

CLIMATE.

In speaking or writing of the climate of a country, it has become the established custom to allude to it as the "finest in the world," and draw a comparison with the "glorious skies of sunny Italy." Most generally those comparisons are far-fetched, and have no real existence except in the writer's imagination. Arizona needs no such fictitious aids to enhance the beauty of its climate. She can show as bright skies, as pure air, as bracing an atmosphere, as lovely, cloudless days, as brilliant starlit nights, as that land over which poets and painters have raved, and sane people have gone into ecstasies. The climate of Arizona suits all constitutions. In the south it is warm and dry, while the elevated plateaus of the north possess a cool, bracing temperature, well adapted to persons who have lived in northern latitudes. The winter in the southern portion of the Territory, and especially at Yuma, is perfection itself. Speaking of the latter place, the celebrated traveler, Ross Browne, has said: "The climate in winter is finer than that of Italy. It would scarce be possible to suggest an improvement. I never experienced such exquisite Christmas weather as we enjoyed during our sojourn." This portion of the Territory is fast coming into favor as a sanitarium for those troubled with pulmonary diseases. The purity, dryness, and elasticity of the air make it unequaled on the continent for the cure of consumption, kidney diseases, and rheumatism. While the heat in summer is high, its peculiar dryness prevents any injurious effects, and sunstrokes are rarely heard of in Arizona.

There is no climate so conducive to longevity. This is attested by the great age reached by Mexicans and Indians born and bred here. Centenarians are not uncommon among these people, and there are many of them who have passed the one-hundred milestone. Barred by the peninsular continuation of the Sierra Nevada from the north-west trade winds, Arizona has to depend for moisture on the winter snows that fall in the northern part of the Territory, and the summer rains that are borne hither on the wings of the south-west trade winds. These cloud-bearing winds, after sweeping over northern Mexico, reach Arizona about the first of July, when the rainy season commences, and last until the middle of September. With the coming of those rains, the summer proper of Arizona begins; grass and vegetation spring up as if by magic, flowers cover the valleys, plains, mesas, and mountain sides, and all nature rejoices at the watery dispensation. In the mountains of northern Arizona the snowfall sometimes reaches a depth of four or five feet. It rapidly disappears from the plains and valleys, but on some of the lofty mountain peaks, like the San Francisco, it remains until the middle of summer. During the snowfall in the upper regions, the plains and valleys of central and southern Arizona are blessed with copious showers. The spring, though dry, is one of the most delightful seasons of the year. In the northern part of the Territory, vegetation takes a rapid start from the moisture caused by the winter snows, grass becomes green, and continues until the summer rains bring forth the vigorous growth of rich grammas. The winter climate of Tucson, Tombstone, Florence, Phœnix, and other points in the south, partakes of the character of Yuma; the mild, balmy air, the days with their clear, cloudless skies, and the nights brilliant with countless stars, like diamonds set in an azure field, make living during the winter months in Southern Arizona a luxury found but in few spots on earth. The winter in the northern portion of the Territory has that cool, bracing quality found in elevated regions; its spring and summer are delightful, the nights are cool and pleasant, making a pair of blankets a comfortable auxiliary to a good night's rest. It would be difficult to find anywhere a climate which possesses the golden mean—not too cold in winter nor too warm in summer—of the plateau of Northern Arizona. As a summer resort the pine-clad mountains of Yavapai and Apache counties, with their springs of clear, cold water, and beautiful, grassy valleys, are not excelled by any portion of the American Union.

Epidemic diseases are unknown in Arizona. Along some of the water-courses in the southern part of the Territory, chills and fever of a mild type prevails during the months of August and September, but is easily broken. It can be truthfully said, that no country possesses a healthier or more uniform climate. The air is dry, pure, exhilarating; there is health in every breeze, and vigor, long life, strength, and happiness under its glorious skies. Those who are suffering from pulmonary complaints or rheumatic affections will find in this favored clime the balmy air and the healing qualities to build up their shattered constitutions. As showing the temperature at different points throughout the Territory and the rainfall for a year, the following tables, kindly furnished by the Signal Service bureau, are appended.

The city of Tucson is 2,500 feet above sea level. The annexed table gives the maximum, minimum, and mean temperature for a year, together with the rainfall.

1880.	Maximum.	Minimum.	Difference.	Amount of Rain or Melted Snow (Inches).
January	78.0	14.0	64.0	0.56
February	77.0	20.0	57.0	0.15
March	87.0	35.0	52.0	0.41
April	88.0	36.0	52.0	0.04
May	104.0	44.0	60.0	0.00
June	110.0	60.0	50.0	0.00
July	108.0	65.0	43.0	1.62
August	106.0	66.0	40.0	1.28
September	106.0	58.0	48.0	1.89
October	94.0	40.0	54.0	0.09
November	73.0	30.0	43.0	0.00
December	80.0	28.0	52.0	0.57
Annual means	92.6	41.3	51.2	0.55

*Temperature at Fort Yuma, from March, 1880, to March, 1881.
The fort is 267 feet above sea level.*

MONTH.	Maximum.	Minimum.	Mean.
1880.			
March...........................	73.97	37.80	55.88
April...........................	83.40	45.10	64.25
May............................	95.13	53.45	74.29
June............................	103.53	64.40	83.96
July............................	105.26	70.74	88.00
August..........................	106.42	70.42	88.42
September.......................	100.00	66.43	83.21
October.........................	91.19	51.93	71.56
November........................	73.50	36.10	54.80
December........................	69.42	37.35	53.38
1881.			
January.........................	67.42	32.09	49.75
February........................	78.46	39.07	58.76
March...........................	79.70	39.61	59.65

*Table showing monthly means of thermometer, amount of rainfall,
and maximum and minimum thermometer, for the year ending
June 30, 1881, at Prescott, 5,600 feet above sea level.*

MONTH.	Total Rainfall or Melted Snow (Inches).	Monthly Mean Thermometer.	Maximum.	Minimum.
1880.				
July............................	2.34	72.6	92	45
August..........................	2.80	71.4	92	40
September.......................	1.26	64.4	90	29
October.........................	0.18	52.3	77	48
November........................	0.42	36.3	65	—1
December........................	1.84	37.8	63	11
1881.				
January.........................	0.16	34.7	62	5
February........................	0.10	40.8	76	10
March...........................	2.91	49.2	78	0
April...........................	0.67	56.8	82	26
May.............................	0.44	62.2	89	33
June............................	0.00	71.3	96	38
	13.12	54.1	96°	—1

The records for only five months of the present year are available from Camp Grant, situated in an elevated region, nearly 5,000 feet above the level of the sea. The climate is among the most delightful in the Territory:

1881.	Mean.	Maximum.	Minimum.
January.........................	41.23	56.12	30.23
February........................	48.09	67.82	37.17
March...........................	50.06	62.03	39.05
April...........................	62.98	78.80	47.10
May.............................	70.26	83.77	54.64

Mean, maximum, and minimum temperature, and amount of rainfall at Fort Mohave, A. T., during the twelve months commencing July 1, 1880, and ending June 30, 1881, rendered by A. A. Surgeon John F. Minor, U. S. A.

MONTHS AND YEARS.	Temperature.			Rainfall (Inches).
	Mean.	Maximum.	Minimum.	
1880.				
July.	91	111	67	....
August	89	109	63	.81
September	82	105	58	.07
October	70	94	50	....
November	52	85	28	....
December	53	70	34	.38
1881.				
January	49	72	30	....
February	59	82	35	....
March	61	96	35	.75
April	74	98	56	.71
May	79	101	62	.01
June	86	108	68	....

This camp is in latitude 35° 24', and longitude 114° 34' west from Greenwich, and is 600 feet above the sea level. It is in the valley of the Colorado, and is considered one of the hottest places on the globe.

RAILROADS, TELEGRAPH AND STAGE LINES.

RAILROADS.

The completion of the Southern Pacific railroad across Arizona marks a new era in the history of the Territory. No longer is it an unknown land, isolated from the busy centers of civilization, trade, and active industry; the dangers and discomforts of long and dreary stage rides, have been superseded by the luxury of the palace car, and a trip to the "marvelous country," at the present time, will be found both pleasant and profitable. The Southern Pacific enters Arizona at Yuma and crosses the Territory between the thirty-second and thirty-third degrees of latitude. Its length within the boundaries of Arizona is over 400 miles. Since the building of the road, many towns and mining camps have sprung up in the country adjacent; an army of prospectors, traders, and speculators has filled the southern counties, and the steadily increasing volume of bullion which is finding its way out of the country, is an earnest of what other portions of the Territory will do when they are likewise in possession of rail communication. At Deming, in New Mexico, about 90 miles east of the Arizona line, another great transcontinental route, the Atchison, Topeka, and Santa Fe railroad, forms a junction with the Southern

Pacific. This line (Atchison, Topeka, and Santa Fe) begins at Kansas City, Missouri, traverses the plains of Kansas and Colorado, enters New Mexico, and passes down the Rio Grande valley, from whence the main line turns west towards Arizona, while another branch follows the Rio Grande to El Paso. From Deming, the Atchison, Topeka and Santa Fe Company have their road surveyed to Tombstone and Tucson, where it is expected it will connect with the branch which is now building from Guaymas, through the State of Sonora. The opening of this great thoroughfare will give Southern Arizona direct rail connection with the Gulf of California, as its junction at Deming with the Southern Pacific has already linked it with the Mississippi valley and the Atlantic seaboard. Among the branch roads projected from the line of the Southern Pacific, is that from Benson station to the city of Tombstone, a distance of twenty-eight miles. Ground has been broken for this branch, and it will be finished at an early day. A branch has also been surveyed from Wilcox to the town of Globe. The length of this proposed line will be something over 100 miles. It will pass through one of the best grazing portions of the Territory, by the lately discovered coal-fields near the Gila, and will open up to capital and imigration that rich mineral region which has Globe for its center.

Another branch line is in contemplation from Casa Grande station to Pinal, by way of Florence. It will pass through the rich valley of the Gila and penetrate the extensive mineral region embraced in the Pioneer, Pinal, Mineral creek, and other rich districts of Pinal county.

The Southern Pacific company have surveyed a line from Yuma to Point Isabel, on the Gulf of California. A good harbor is said to exist at that place. The building of this branch will give the Territory another outlet to tidewater on the gulf. A line has also been surveyed from Yuma to the rich mining camps of Castle Dome and Silver district, on the Colorado river.

In the northern part of the Territory, the construction of the Atlantic and Pacific railroad is making rapid progress. This road leaves the Atchison, Topeka, and Santa Fe at Albuquerque, and takes a westward course across the Territory, following nearly the thirty-fifth parallel of north latitude. The road will pass about 50 miles north of Prescott, the capital of Arizona, and will cross the Rio Colorado at the Needles. This road will have termini at San Francisco and San Diego. The Atlantic and Pacific line will open to the capitalist, the miner, and the stock raiser, some of the finest grazing and richest mineral regions to be found on the continent; it will also pass through the best-timbered portion of the Territory. A franchise has been granted by the last Legislature to build a branch from Prescott to the Atlantic and Pacific. The distance, as has been before stated, will be about 50 miles, most of the way over a smooth, rolling country. The extensive mining, farming, and grazing interests, of which Prescott is the natural

center, require the construction of such a road, and it will no
doubt be completed within a short time.

Besides the roads now building and those projected, which
have been mentioned, the Utah Southern is being pushed
down to the Colorado river, with the intention, as is generally
supposed, of seeking an outlet on the Gulf of California. This
would give Arizona a connection with the Union Pacific and
another route to the East and West.

From this brief review of the railroad situation, it will be
seen that all the principal points in the Territory will soon be
in possession of rail communication. It is safe to say that
within the next two years all the leading towns and mining
camps will be linked to the outside world with iron bands. The
benefits which cheap freights and rapid transit will confer on
the Territory are almost incalculable. Besides that the building
of the road on the thirty-fifth parallel will give the people of
Arizona a competing line to the marts of the East and the West,
it will help to maintain a healthy competition, and prevent
discriminating and oppressive charges on freight and travel
which the corporation controlling the Southern Pacific have
always shown a disposition to indulge in when there was no
opposition.

TELEGRAPH LINES.

The Western Union Telegraph Company have a line through
the Territory along the track of the Southern Pacific railroad,
and. connecting at Yuma, Tucson, and Tombstone, with all
points east and west. The government has a line connecting
all the principal military posts throughout the country.
Branches of this line, which connect with the Western Union,
run to Prescott, Phœnix, Florence, and other towns. It is
under the charge of the Signal Service bureau, is a great con-
venience to the people, and, for years, was their only means of
quick communication with the outer world. From Globe to
the San Carlos Indian reservation, a line has been built by a
stock company composed of prominent citizens of the former
town. At the latter place it connects with the United States
military line. With the completion of the Atlantic and Pacific
railroad, another telegraph wire will stretch across the northern
portion of the Territory, bringing the chief settlements in communication with all parts of the civilized globe.

STAGE LINES.

The mail facilities of Arizona, while not perfect, are better
than are generally found in the remote Territories. Stage lines
connect with the leading towns and mining camps distant from
the railroad, and mails are carried with regularity and dispatch.
The opening of the Southern Pacific has brought the Territory
in close connection with the East and West; letters from New
York reach Tucson within six days, while Prescott is only four
days distant from San Francisco. All the principal towns are
supplied with daily mails, while every farming settlement or
mining camp, of any size, has at least a weekly.

From Tucson stages run to Arivaca daily, connecting with the mining camps adjacent. This well-appointed line carries the mails to Altar and other points in Sonora. Another daily line runs from Tucson to Hermosillo, by way of Calabasas. From Tucson to Silver Bell, a flourishing mining camp, 50 miles distant, there is a semi-weekly line.

From Tombstone to Benson, on the Southern Pacific railroad, there are two daily lines of six-horse coaches, carrying mails and passengers. They have good stock, and make fast time. A tri-weekly mail is carried from Tombstone to Harshaw, passing by Camp Huachuca. A daily line is also run from Tombstone to Charleston, and a tri-weekly to Bisbee. There is a daily line from San Simon, on the Southern Pacific railroad, to the prosperous mining camp of Galeyville, in the Chiricahua mountains. From Wilcox station, daily mails are carried to Safford, the county seat of Graham county, and also to Globe, the county seat of Gila. This line passes by Camp Grant and San Carlos.

A daily stage connects Casa Grande with Florence. From Florence a line runs to Globe, by way of Riverside, and another daily stage carries mails and passengers to Pinal and Silver King. This company have good stock and comfortable coaches.

Phœnix is connected by a daily line of coaches with the railroad at Maricopa, and by a daily and tri-weekly line with Prescott. A tri-weekly mail is also carried to Fort McDowell.

Prescott, distant 140 miles from the Southern Pacific at Maricopa, has one daily and one tri-weekly line of coaches to that point. These stages pass through Phœnix, and passengers have the choice of two routes to Northern Arizona from the south—by way of Wickenburg, and by way of Black Canyon. Good stock and roomy coaches are run on these lines. A tri-weekly line runs from Prescott to Mineral Park, the county seat of Mohave county, and also to Alexandria, a mining camp 30 miles south. A new line has been established from Prescott to the terminus of the Atlantic and Pacific railroad, which will be increased to a daily, as the road advances westward.

Mohave county has a tri-weekly mail from Mineral Park and Cerbat to Prescott; there is also a tri-weekly line to Port Mohave, on the Colorado river.

St. Johns, the county seat of Apache county, has regular mail connection with the Atlantic and Pacific railroad, and with the southern portion of the Territory. Yuma has a tri-weekly mail line to Castle Dome, Silver District, and Ehrenberg. Nearly all these lines have comfortable coaches and good stock. Passengers will find eating stations at convenient distances. The traveling is nearly all by day, and no pleasanter trip can be imagined than a ride on the outside seat of a Concord coach, behind a good team, over the ever-changing panorama of mountain, valley, and table land which make up the bold outlines and wonderful perspective of Arizona scenery.

7

THE INDIAN TRIBES.

No description of the Territory would be complete without some account of its Indian tribes. For years the name Arizona was indissolubly linked with savage massacres, fiendish murders, and sickening tortures; it was the "dark and bloody ground" of the frontier, where the few whites who had the temerity to penetrate, carried their lives in their hands, went armed to the teeth, and kept constant watch for the treacherous foe. Perhaps no portion of the American continent has witnessed a more deadly struggle than that waged by the pioneers of Arizona against the murderous Apache.

For nearly fifteen years this warfare was maintained by the handful of whites scattered over the Territory from the Utah boundary to the Sonora line. Isolated from the centers of population, and surrounded on all sides by their savage foes, the gallant band maintained the unequal contest, and although. hundreds of them fell victims to savage treachery, and left their bones to bleach on the desert plain and mountain side, the red man was compelled at last to yield to his destiny. A volume would be required to give an account of the long and bloody struggle, of the lonely ambush, the midnight attack, the hand-to-hand encounter, the shrieks of women and the cries of children, the flames of burning dwellings, and the fiendish yells of the infuriated savages. No writer of Indian fiction ever imagined more desperate combats, more hair-breadth escapes, more daring courage and self-sacrificing devotion, than the history of the Apache wars in Arizona will show when they are fully written. The savages were at last conquered by General Crook and the gallant officers and men under his command, in 1874, and placed on reservations, where they still remain.

The San Carlos reservation is situated in the eastern part of the Territory, and embraces portions of Gila, Graham, and Apache counties. It is a well-watered region, and has some of the finest farming land in Arizona. It contains at present 4,979 Indians, divided into the following bands: White Mountain, Chiricahua, Coyoteros, San Carlos, Aguas Calientes, Mohaves, Yumas and Tontos. With the exception of the Yumas and Mohaves, all the Indians on this reserve belong to the Apache family. There are 15,000 acres of land within the limits of the agency which can be irrigated; about 1,000 acres have been brought under cultivation, and 250,000 pounds of barley, 5,000 pounds of wheat, and nearly 800,000 pounds of corn have been raised by the aboriginal agriculturists the present year. A large school-house has been built and fitted up with dormitories, dining-room, bath-rooms, etc., where 30 scholars, all boys, receive board and tuition. The reservation is in charge of an agent, with the following assistants: Clerk, storekeeper, physician, chief of scouts, blacksmith, carpenter, three butchers, three teamsters, and two interpreters. The Apaches at this reservation were once the most formidable foes of the whites, and the Chiricahuas, led by the famous chieftain Cachise, were long the terror of Southern Arizona, and have marked every mile of the road from the Rio Grande to Tucson with the graves of their victims. The Apaches, as far back as the history of the Territory extends, were always at war with their neighbors; lived by murder, robbery, and rapine; their hand was against every man, and every man's hand was against them. They kept the Pimas, Moquis, Papagoes, and other semi-civilized tribes continually on the defensive, and it has been supposed that they were the destroyers of the ancient civilization which once flourished in this Territory.

The tribe is divided into sub-tribes, and the sub-tribes again into bands, governed by petty chiefs or captains. In their civil polity they are republicans, pure and simple. The chief or head man is elected by the popular voice, and when his course becomes obnoxious to the majority, he is removed and another chosen in his place. These Indians are polygamists, and keep as many wives as their fancy may dictate, or as they can induce to live with them; they indulge in no marriage ceremony, but the bridegroom is expected to make a present to the bride's father, when he carries her off from the parental *wickiup*. The women are the hewers of wood and the drawers of water, the Apache braves, like all other Indians, considering it a degradation to work. Since their removal to the reservation, however, many of them have laid aside their pride, and plied the shovel and the hoe with commendable vigor. Their moral condition is like that of all other Indians who have been brought in contact with the whites. In their wild state, infidelity on the part of the wife was punished by cutting off the nose, but since their intercourse with the pale faces, they have adopted a less severe code. All the Apaches are inclined to spiritualism, and are very superstitious; they also believe in witches and have almost implicit faith in their medicine men; are cremationists, and burn their dead. Their habits are filthy; they have adopted many of the white man's vices, and none of his virtues; whisky and civilization are too much for them; the once warlike and powerful tribe of the Apaches are gradually passing away, and the land of which they were once the absolute lords and masters, will, in a short time, know them no more forever.

The Pima and Maricopa tribes have a reservation on the Gila river, commencing about nine miles below Florence and extending down the stream for nearly thirty-five miles. The Maricopas were once a part of the Yuma tribe, but in the middle of the last century they allied themselves with the Pimas, and they have ever since lived together in peace and harmony, although their manners, customs, laws, religious ceremonies and language are as distinct as if they were thousands of miles apart. The tribes number about 5000, 500 being Maricopas. They live in small villages; the houses are built by placing poles ten or twelve feet long in a circle of about twelve feet in diameter at the bottom, and fastened together at the top. These poles are then covered with grass and mud, only a small opening being left for a door. Each village is ruled by a chief, who is subordinate to the chieftain of the tribe. All disputes between the inhabitants of the same village are submitted to a council of the old men for settlement, and their decision, be what it may, is final; in disputes between residents of different villages, representatives from all the hamlets are called by the chief of the tribe to settle the differences. They are polygamists to a certain extent, and an annual feast and dance called the *Tizwin* feast, is held in the early summer, when all who so desire, make their choice of mates for the ensuing year. The Maricopas are cremationists, while the Pimas bury their dead.

Besides their reservation on the Gila, a large tract on the north side of Salt river was set aside for their use by an executive order dated July 14, 1878. They cultivate about 400 acres on Salt river, and on the Gila something like 800. Their wheat crop averages about 2,000,000 pounds a year, and is much superior to that of the whites, both in cleanliness and quality. Corn, beans, pumpkins, and sorghum are also raised in large quantities. Living down the Gila, below the mouth of the Salt, there are about 400 Papagoes who cultivate nearly 400 acres. All of these tribes have some cattle and a great number of ponies. The agent for the Pimas and Maricopas resides at Sacaton, on the Gila, and distributes the government annuities among them. Two schools have been established at this point, with what success we have not learned. These Indians are peaceable and industrious; besides their farming they manufacture *ollas*, baskets, and formerly made some fine blankets. Many of them, by their industry and thrift, have accumulated property to the value of several thousand dollars. They have ever been the friends of the whites, and during the Apache wars their doors were always open for the unfortunate American hard pressed by the foe.

The Pimas were settled on their present abode when found by the Spanish explorers, nearly 350 years ago. Then, as now, they cultivated the soil, and manufactured earthen vessels, and cotton and woolen fabrics. Their farming is done in primitive style, using wooden plows, and threshing the grain by spreading it in a circle on the earthen floor, and driving a band of ponies over it. The Pimas are good warriors, and for centuries resisted successfully the attacks of their hereditary enemies, the Apaches. They have great faith in their medicine men—so long as they are successful in effecting cures. Repeated failures, however, are apt to lead to serious consequences. A case has lately occurred where an unfortunate follower of Galen, having sent three patients, in succession, to the happy hunting-grounds, was taken by a strong guard to the cemetery near Phœnix, and summarily dealt with by having his brains knocked out with a club. If civilization should adopt such a plan, what a thinning out there would be in the medical profession!

The Papagoes were partly civilized when discovered by the Spaniards, over three centuries ago. They were converted to Christianity by the early Catholic missionaries, and still remain steadfastly attached to that faith. Of all the Indians of the

Territory, they are the most industrious, virtuous, temperate, and thrifty. They live by cultivating the soil, and by stock-raising. They have always been peaceable and well-disposed, and during their long contest with the Apaches, they rendered valuable services to the whites. They have never asked or received assistance from the government, although no tribe has so well deserved it. They speak the same language as the Pimas, and are supposed to be a branch of that tribe; but, unlike them, they cut their hair, wear hats, and dress after the fashion of the lower classes of Mexicans. Many of them are employed by the farmers of the Gila and Salt-river valleys, during the harvest season, and have proven steady and faithful laborers. The tribe numbers about 6,000. They have a reservation on the Santa Cruz, south of Tucson, where they raise considerable wheat, barley, corn, pumpkins, melons, etc., and a great many cattle and horses. Their location is a good one, being well watered and timbered, and containing some of the finest land in the Territory. A number of them still live in their old home, the Papagueria, south-west of Tucson, engaged principally in stock-raising. The Papagoes are in charge of the agent at Sacaton. A school is maintained for their benefit, at San Xavier, by the Sisters of St. Joseph, and is largely attended.

The Colorado River reservation was established by act of Congress, March 3, 1865. Since then it has been enlarged, and contains at the present time about 140 square miles, situated between Ehrenberg and La Paz, with a total Indian population of 1,010, composed of the following tribes: Chim-e-hue-vis, 208; Mohaves, 802. Besides the agent in charge, there is a physician, clerk, farmer, carpenter, blacksmith, teacher, matron, and cook. It is said that the morals of these Indians are better than could have been expected from their lax marriage rules; "prostitution is not universal by any means, and is confined to a few depraved women of the tribes." The Indians on this reservation cultivate small patches of ground along the Colorado, raising corn, wheat, melons, pumpkins, etc. The government has expended large sums in opening irrigating canals, and it is hoped that they may soon become self-sustaining. They were once in active hostility against the whites, but the crushing defeat they received at the hands of Colonel Hoffman, in 1859, completely broke their spirit, and they have never since shown any disposition to go on the war-path.

The Yumas live on the Colorado river, ranging from Yuma down towards the gulf. They raise some corn and vegetables on the Colorado bottoms, but spend most of their time loafing around the streets of the town, doing small jobs and carrying messages for the whites. They were once a powerful tribe, but intemperance and immorality have done their work upon them, and they are now the lowest and most debased of all the Indians in the Territory.

The Hualapais live in the mountains of Mohave county. They are a brave and warlike race, and gave the early settlers a great deal of trouble. They were placed on the Colorado reservation, but the enervating climate of the river bottoms was fatal to Indians accustomed to the purer air of the more elevated regions, and they were allowed to return to their native hills. They are industrious, and many of them find employment at the settlements and mining camps throughout the county. They are generally self-supporting, though the government occasionally issues them supplies. The Hualapais did good service during the Apache wars, several companies enlisting as scouts, and fighting bravely by the side of the troops. They have become debased by their intercourse with the whites, and are rapidly decreasing. They number about 700, divided into bands.

The Ava-Supies live in the deep canyon of Cataract creek, a tributary of the Colorado, which rises in Bill Williams mountain, north of Prescott. The band numbers about 300 men, women and children. The narrow valley in which they live averages from 100 to 400 yards wide, with walls of sandstone from 2,000 to 4,000 feet, rising perpendicularly on either side. Down in this beautiful glen the climate is almost perpetual summer; and while the icy winds sweep over the elevated plateau, the lovely vale below sees the flowers bloom and the grass green all the year round. Through the center of this valley runs a clear stream; the soil is rich and easily cultivated, producing grain and vegetables of all kinds, also fine peaches and other fruits. A trail leads down the sides of the perpendicular cliffs, from three to six feet wide, and requires a steady nerve to pass over it in safety. Thus, literally shut out from the world, the Supies live in their beautiful canyon, blessed with everything to supply their few and simple wants. They do a large trade in buckskins and dried fruits with the Hualapais, Moquis, and other Indians. They are peaceful, industrious, and contented, and warmly attached to their homes; are kind and hospitable to strangers, and are, in all respects, the most remarkable tribe in the Territory.

The Moquis occupy several villages in the north-eastern portion of the Territory. Their "pueblos" are situated on rocky cliffs from three to six hundred feet above the level of the surrounding plain. On one of these isolated *mesas* are located four of their villages. Three other villages occupy as many rocky bluffs or *mesas*. The houses are of stone, and built in terraces, in such a manner that to enter the lower story it is necessary to climb to the top and then descend. The inhabitants of Oraybe, west from the Moquis, are of different origin and language, although their manners, customs, and mode of life are the same. Water is brought to these pueblos, perched on those rocky crags, from a half to two miles distant. The valley below, although sandy and · barren-looking, produces good crops of corn, pumpkins, melons, and fine peaches. About three thousand acres are in cultivation at the different villages. They have large flocks of sheep and goats, which they carefully guard from the raids of their more warlike neighbors, the Navajos. The Moquis are temperate, industrious, and true to their marriage relations. They make blankets, baskets, and *ollas ;* have lived in their present abode since we have any knowledge of them, and are the same in all respects to-day as they were three hundred and forty years ago, when Coronado and his followers, in their search for the Seven Cities of Cibola, first met them. An agent has been appointed for them, and a boarding-school established, which is proving a gratifying success.

The Navajo reservation is located in the north-eastern corner of the Territory, adjoining the line of New Mexico, and embraces an area of 5,200 square miles, the greater portion being fine grazing land. The Navajos are the main branch of the Apache family, and are probably the most intelligent, active and enterprising of all the Indians in Arizona. Their manufacture of fine blankets has long been admired, and in their agricultural and pastoral possessions, they are one of the richest tribes in the United States. They own about 15,000 fine horses, over 400,-000 head of sheep, nearly 2,000 head of cattle, besides mules, *burros*, etc. They derive over $30,000 annually from the sale of blankets, sashes, etc. Every family has its loom, where the women are constantly employed. The Navajos are a warlike race, have long kept their Moquis and Zuni neighbors in wholesome dread, and at one time were the terror of the Rio Grande valley. Since their subjugation by the government in 1860, they have made rapid strides in prosperity, and are said to be the only Indians who are increasing. They number at present about 15,000. Their agency is established at Fort Defiance.

The total number of Indians in the Territory is about 25,000. The power of the wild Apache has been broken, and he no longer obstructs the path of progress and civilization. The Indian question in Arizona has been settled forever; the wild tribes are fast passing away, and in a few years will have entirely disappeared, leaving behind only a name linked with bloody deeds and savage atrocity.

MISCELLANEOUS.

WAGES AND COST OF LIVING, SOCIETY, MANUFACTURES.

People who are looking for homes in a new country, naturally feel an interest in knowing the rates of wages paid, and the cost of living in the region to which they think of emigrating. In this chapter we shall endeavor to answer the many inquiries which are being made from the East, and from the Pacific States and Territories, asking for information on these points.

Miners are paid $4 per day throughout the Territory. This is the rate of wages for underground work which has prevailed in the neighboring State of Nevada, and which has been established in Arizona. In some small and isolated camps a lower rate has obtained, but good workmen, who understand their calling, can not be hired for less than the prevailing rates.

Blacksmiths receive from $4 to $6 per day, first-class workmen commanding the latter price. Carpenters get from $4 to $5 per day; bricklayers and masons from $5 to $6 per day; engineers from $5 to $6 per day; printers from $4 to $5 per day; clerks from $50 to $100 per month and board; teamsters from $40 to $70 per month and board; herders from $30 to $40 per month with board; farm laborers from $30 to $40 per month; and day laborers from $2 50 to $3 50 per day.

The supply of labor is generally in excess of the demand. Like all mining countries which have received a sudden impetus from the opening of railroads, Arizona has drawn within its borders a number of people who have found themselves, on their arrival in the country, destitute of means. While there is always a chance for men of energy and industry to make their way, it is not advisable for mechanics and laboring men, who have no means, to rush to Arizona. While those who are employed obtain good wages, it must be borne in mind that this is a country whose many resources are just beginning to be developed, and that the demand for labor is limited. To men who have some means; who are in a position to take advantage of the many profitable openings that present themselves; who may be in possession of a small capital to begin the battle of life; who have the wherewithal to try their fortune in seeking for the treasures that lie hidden in our mountain fastnesses, Arizona offers advantages not equaled by any State or Territory in the Union. But of the workingman, who has only means sufficient to bring him to the country, and is dependent solely on his daily labor, Arizona has already enough, and it is not the desire or intention of this publication to hold out uncertain inducements to that class of emigrants.

The cost of living in the Territory is not more expensive than could be expected in a country, the greater portion of whose supplies are brought from such a distance. With the exception of some grain, flour, hay, and vegetables, everything worn or consumed by the people of Arizona is shipped from California or the East. In Tucson board can be had at from $6 to $8 per week, and at the leading hotels at from $1 to $2 50 per day. In Tombstone, board is from $8 to $10 per week, and in the different mining camps throughout the southern portion of the Territory, the same rates prevail. Rents in Tucson and Tombstone are not high, considering the rush of emigration to those towns, and the remarkable advance in real estate. A comfortable residence of three or four rooms, in a suitable location, can be had in Tucson at from $20 to $30 per month. The rates are about the same in Tombstone. Clothing, boots and shoes, dry goods, groceries, and everything necessary for housekeeping, are sold at fair prices. A suit of clothing can be bought at from $15 to $30; a pair of boots at from $4 to $8, and all other articles in a like proportion. Of groceries, sugar is 20 cents per pound; coffee, 25 cents; flour, $5 per cwt.; beef, 8 to 12 cents per pound; and vegetables and all other articles of food at similar rates. In Phœnix, the agricultural center of the Territory, prices of clothing and groceries are about the same as in Tucson and Tombstone, while grain, flour, vegetables and fruits, are much cheaper.

In Prescott and throughout Northern Arizona, the rates of wages do not differ materially from those which exist in the southern country. Board in Prescott is from $8 to $10 per week. Groceries, clothing, and provisions are a trifle higher than in Tucson and Tombstone, owing to the greater distance from the Southern Pacific railroad, and the increased charges on freight. The opening of the Atlantic and Pacific will give Northern Arizona a direct line to the markets of the East, and supplies and material of all kinds can be laid down at Prescott and the northern mining camps, at much lower rates than at present. From this brief summary it will be seen that the cost of living in Arizona, taking into consideration the long distances from the sources of supply, can not be considered high; and although these figures may appear rather large to people accustomed to those prevailing toward the rising sun, it must be borne in mind that every branch of labor, and every profession or calling, receives a just and generous remuneration for its services in this prosperous and progressive Territory of the South-west.

SOCIETY.

There is no Territory on the distant frontier where law and order are so strictly maintained, or where the rougher elements, peculiar to the border, observe so mild-mannered an attitude, as in Arizona. In the newest mining camp, as well as in the larger towns, like Tucson and Prescott, life and property are as secure as in older communities who boast of their culture and civilization; and if sometimes the festive "cowboy" from Texas, or the "bad man from Bodie," should forget himself while under the influence of "fighting" whisky, he is quickly brought to a realizing sense of the situation by the strong arm of the law. On the opening of the Southern Pacific railroad, a crowd of outlaws from the East and the West flocked into Arizona, but the prompt and energetic action of officers and citizens, soon compelled that gentry to seek fresh fields. Even the contests over mines, which seem to be inseparable from a "live" camp, have been fewer than in most of the mineral States and Territories; the pistol and the shotgun have been laid aside, and the law allowed to have its course.

Tucson, Tombstone, Phœnix, and Prescott are incorporated under the laws of the Territory. They have an efficient police force, and the best of order is maintained. In fact, it has been remarked by travelers and new-comers that Arizona has less of that typical western lawlessness than any region they had visited on the frontier. In the leading towns of the Territory will be found a society whose culture, intelligence, and refinement will compare with any portion of the Union. Surrounded by churches, schools, newspapers, and the other adjuncts of modern progress, the people of Arizona are among the most intelligent, liberal, and progressive to be found in the United States. The emigrant who decides to cast his lot here will find the foundations of a broad and enlightened society firmly established; he will meet a generous, progressive, and liberal-minded people, ready to lend a helping hand to the new-comer; and he will find order, security, law, and enlightened public opinion ruling the country from the Utah line to the Sonora border.

MANUFACTURES.

The manufacturing interests of Arizona are yet in an embryo condition. Lumber and flour are its chief products at the present time. Yavapai county has three saw-mills near Prescott, and one on the line of the Atlantic and Pacific railroad. These mills turn out a good quality of pine lumber, and supply a large area. Lumber is worth from $20 to $30 per thousand at the mills. At Prescott, there is a sash, door and blind factory, which is kept steadily at work. A small foundry has been established here, but it is now closed.

Maricopa county manufactures nearly three fourths of all the flour produced in the Territory. It has four flour-mills in active operation; one at Phœnix, one three miles east of Phœnix, one on the Grand canal, and one at Tempe. All these mills are supplied with the best machinery and the latest improvements, and turn out a quality of flour preferred by some to the best California. An ice factory has been established at Phœnix which supplies its citizens with a luxury which is almost a necessity during the sultry summer months. Large quantities of sorghum are also manufactured in the Salt-river valley. It is a superior article and finds a ready sale.

The manufacturing industries of Pima county consist of two flour-mills in Tucson, well-appointed establishments, which produce a superior article. A foundry and machine shop was established here in 1880, and is prepared to make every variety of quartz-mill machinery and castings in iron and brass. Several large blacksmith and wagon shops are also in full operation in Tucson, and turn out superior work in their line.

Cochise county has five saw-mills in operation, three in the Huachuca mountains west of Tombstone, and two in the Chiricahua range east of that point. These mills produce an excellent quality of pine lumber, which finds a ready sale in the bonanza camp and the mines adjacent. Tombstone has also a foundry where castings for quartz-mills of every description are manufactured.

The manufactures of Gila are confined to two saw-mills in the Pinal mountains, which supply Globe and the mining camps throughout the county with a superior article of pine lumber. At Yuma is situated the largest wagon factory in the Territory. The peculiar dryness of the climate at this point seasons the wood so thoroughly that it never shrinks. The mesquite, which grows in such profusion on the Gila and Colorado bottoms, makes the very best wagon timber, and the work turned out at this place is considered the most durable and best adapted to the climate of the Territory.

Apache county has several saw-mills steadily at work on the magnificent pines which crown her mountain ranges. Two flouring-mills have been put up on the Colorado Chiquito, which produce a fine article of the staff of life.

In Pinal county there are two flour mills, on the Gila, below Florence, which find profitable employment in handling the fine wheat for which that valley is celebrated. Graham county has two flour-mills in operation at Solomonville. These are about the only manufacturers now in existence in the Territory. That there is here an extensive and a profitable field for the investment of capital in this branch of industry, admits of no doubt. No better opening can be found on the Pacific coast for a woolen factory. The wool is here in abundance; the water-power is here, and the demand, already sufficient to make the venture a paying one, is steadily on the increase. A tannery would be a lucrative enterprise; thousands of hides are now shipped out of the country every year, which should be turned into leather at home. Every broom used in the territory is made abroad, when it has been demonstrated that broom-corn of an excellent quality can be grown in the valleys of the Gila and the Salt rivers. The manufacture of soap is also an enterprise which offers quick returns to any one who will engage in it. Ropes, cloth, and paper of a superior quality have been made from the fibers of the mescal plant, and as the supply is unlimited, there is no reason why a venture of this kind should not be successful. For the man who will be the first to inaugurate some of the manufacturing enterprises alluded to, success is certain. The population is steadily increasing, and the demand for the articles mentioned is increasing in the same ratio. The raw material is at hand, and it only requires capital, energy, and enterprise to reap this virgin field and glean a golden harvest. There are many other industries of a kindred nature to those we have set forth, which can be profitably engaged in, but enough has been said to convince business men of the splendid opportunities which Arizona offers for the successful prosecution of manufacturing industries.

POPULATION—CIVIL AND MILITARY.

According to the census of 1880, Arizona has a population of 41,580, distributed as follows:

Pima county	19,934
Maricopa county	5,689
Yavapai county	5,014
Apache county	3,498
Pinal county	3,040
Yuma county	3,215
Mohave county	1,190

This population is classified as follows:

Whites	35,330
Mulattoes	17
Blacks	87
Chinese	1,601
Indians	4,545
Making a grand total of	41,580

This estimate does not include Indians on reservations and those who live in pueblos. The population of the Territory has rapidly increased during the past two years. The large emigration which the building of the Southern Pacific railroad has drawn to the southern portion of the Territory, shows no signs of slackening. The completion of the Atlantic and Pacific road through Northern Arizona will no doubt attract to that region a human tide equally as large as that which has swept over the southern country. It is not too much to expect that Arizona will double its present population within the next two years, and in three or four years from now, have the requisite number of inhabitants to entitle her to admission as a sovereign State of the Union.

The preponderance of males over females is very marked in Arizona, as in all new countries. The opening of railroads, however, will help materially to equalize this difference, and more evenly balance the sexes. What has been said of the inducements which the Territory holds out to men, will apply also to women. In none of the Western Territories is female labor better paid. Women who are not afraid to work, and are willing to cast their lot with the destinies of this young and flourishing Territory, will find many advantageous opportunities, which they can not hope for in the crowded centers of the East.

MILITARY.

Arizona and Southern California constitute a separate military department, with headquarters at Fort Whipple, near Prescott, Brevet Major-General O. B. Wilcox commanding, with the following staff:

First Lieutenant H. L. Haskell, Aid-de-camp.
Second Lieutenant E. F. Wilcox, Aid-de-camp.

Department Staff.

Major Samuel N. Benjamin, Assistant Adjutant-General.
Major A. K. Arnold, Acting Assistant Inspector-General.
Colonel H. C. Hodges, Chief Quartermaster.
Captain Charles P. Eagan, Commissary of Subsistence.
Surgeon A. K. Smith, Medical Director.
Major W. H. Johnson, Paymaster.
First Lieutenant Carl F. Palfrey, Engineer-Officer.

Commanders of Posts.

Fort Apache, in the Sierra Blanco, is garrisoned by two companies of cavalry and two of infantry, commanded by Colonel E. A. Carr.

Fort Bowie, in Apache pass, the former stronghold of Cachise, has two companies of cavalry and is under the command of Captain C. B. McLennan.

Fort Grant, fifty miles north of Tucson, is garrisoned by two companies of cavalry and two of infantry, with Major James Biddle in command.

Camp Huachuca, south-west from Tucson and near the Sonora line, has a garrison of one company of cavalry under the command of Captain T. C. Tupper.

Fort Lowell, nine miles from Tucson, has one company of cavalry, Captain W. A. Rafferty in command.

Fort McDowell, near the junction of the Salt and the Verde rivers, has one company of cavalry and one of infantry, under the command of Captain A. R. Chaffee.

Fort Mohave, on the Colorado river, is garrisoned by one company of infantry under the command of Captain E. C. Woodruff.

Camp Thomas, on the Upper Gila, has a garrison of one company of cavalry and one of infantry, under the command of Major David Perry.

Fort Whipple, the headquarters of the department, is about one mile east of Prescott. It is garrisoned by two companies of infantry, Captain H. C. Egbert, Post Commander.

The number of troops in the department of Arizona is about 1,200, distributed over the entire Territory. No more efficient force is found on the frontier, and no portion of Uncle Sam's domain is more carefully looked after. Too much credit can not be awarded to General Wilcox, and the officers and men under his command, for the manner in which they have guarded the important interests confided to their charge; quelled all symptoms of hostility among the Indians within the Territory, prevented the incursions of hostile bands from abroad, and kept securely the long line of frontier bordering on Mexico. The people of Arizona owe to the army a debt of gratitude which can never be forgotten; their services in subduing the savage Apache, and opening this country to settlement and civilization, will ever be held in grateful remembrance, and will constitute one of the brightest pages in the history of the Territory.

CIVIL.

Arizona, as one of the Territories of the Federal Union, has her leading civil officers appointed by the President. The people have the privilege of electing a delegate to Congress, who has no vote. They are also permitted to elect a Legislature every two years, who enact laws, subject to the approval of Congress. The following is a list of the Federal officers of the Territory at the present time:

Delegate in Congress, Granville H. Oury.
Governor, John C. Fremont.
Chief Justice, C. G. W. French.
Associate Justices, W. H. Stilwell, De Forest Porter.
United States District Attorney, Everett B. Pomroy.
United States Marshal, C. P. Dake.
Surveyor-General, John Wasson.
United States Depositary, C. H. Lord.
Collector of Internal Revenue, Thomas Cordis.
Collector of Customs, W. F. Scott.

ANCIENT RUINS.

The evidences of an ancient civilization which are met with in the ruins scattered throughout the Territory, have long been a subject of earnest inquiry among savants and explorers. The character and extent of these ruins prove conclusively that the region now known as the Territory of Arizona was, at some period in the past, the seat of a civilization much further advanced than that which occupied the land when first discovered by Europeans.

First among these prehistoric relics, both in its extent and state of preservation, is the Casa Grande, about six miles below Florence, in the valley of the Gila, and about five miles south of that stream. This ruin was discovered by Coronado's expedition in 1540. It was then four stories high, with walls, six feet in thickness. Around it were several other ruins, some with the walls yet standing, which have since succumbed to time and the elements. The Pima Indians, who, then as now, were living in the immediate vicinity, had no knowledge of the origin or history of the structure. It had been a ruin as far back as tradition extended in their tribe, and when or by whom erected was as much a mystery to the dusky natives as to their European visitors. Father Pedro Font examined the Casa Grande in 1775, and describes the main building as "an oblong square, facing to the cardinal points of the compass. The exterior wall extends from north to south four hundred and twenty feet, and from east to west two hundred and sixty feet. The interior of the house consists of five halls, the three middle ones being of one size, and the extreme ones longer. The three middle ones are twenty-six feet in length from north to south, and ten feet in breadth from east to west, with walls six feet thick. The two extreme ones measure twelve feet from north to south, and thirty-eight feet from east to west." At present the ruins are about two stories high, and are rapidly crumbling away. The walls are composed of a material looking like concrete or grout. The dimensions of the ruin still standing are about 50 by 30 feet. It is divided into many small rooms, and plastered with a reddish cement. The walls still show small round holes where the rafters had entered, charred pieces of which are yet found imbedded in the adobe. The interior room is the largest, and is still in a fair state of preservation. All around the main building are mounds and traces of ruins, which go to show that a large city existed here at one time. The course of an immense irrigating canal, which watered the plain where the ruins now stand, has been followed to the Gila above Florence, forty miles distant.

Near Tempe, in the Salt-river valley, are found the ruins of extensive buildings, which are supposed to have been even larger than the Casa Grande. The foundations of one have been traced, which measures 275 feet in length and 130 feet in width. Excavations made in these mounds have brought to light several *ollas* which were filled with charred bones. The remains of a large irrigating canal are traced near the ruins. The road from Phœnix to Tempe follows the bed of the ancient water-course for a considerable distance; it is much larger than any in use by the modern occupants of the valley. The ruins of canals and buildings which are yet found in the plain between the Gila and the Salt rivers go to show that this region, now so desolate, was at one time thickly inhabited. At many other points in the Salt-river valley the marks of a civilization which once flourished here and made the desert to smile with industry, are yet plainly traced. All about the ruins are found fragments of pottery, painted in various colors and highly glazed.

In the valley of the Upper Gila, known as Pueblo Viejo, are found extensive mounds similar to those of the Salt river. Traces of buildings, irrigating canals, broken pottery, etc., are met with in every direction. Ruins of a like character are encountered at different points all along the Gila river. On the San Pedro, near its junction with the Gila, are remains of what must have been a large city. The foundations were of stone, laid in a coarse cement. Numerous ruins are found along the Verde and its tributaries, in the Agua Fria valley, and in the mountains and valleys extending for fifty miles in every direction from Prescott. Some of the structures on the Verde and Beaver creek, are among the most interesting in the Territory. On a hill overlooking the river, below Chino valley, is a series of ruins of stone houses; on another hill, about three miles east, are found the remains of many other stone buildings. In the valley of the Verde, traces of its early inhabitants are found in every direction. Opposite Camp Verde are a number of stone ruins, overlooking the river. Two miles down the stream, on an elevated *mesa*, an ancient burial ground has been discovered. On Beaver creek, a tributary of the Verde, are found many interesting cave dwellings. They are walled up in front, and look like the rocky bluffs out of which they have been excavated. Cisterns made of cement, and in a remarkable state of preservation, are found near many of these dwellings. One of these caves is eighty feet across its front, and nearly one hundred feet above the base of the cliff. The interior is divided into many rooms, the height of the roof being about fifty feet. The wall in front is pierced by two loopholes, through which a view of the country for some distance around, can be had.

In Chino valley, twenty miles north of Prescott, are found many interesting stone ruins. Large *ollas*, filled with charred corn and beans, have been unearthed from these mounds. Several skeletons have been discovered, and also a number of stone hammers and axes. There is every reason to believe that the inmates died by violence, the doors and windows being walled up, evidently as a protection against a hostile foe.

In the vicinity of Walnut Grove, twenty-five miles south of Prescott, are found the ruins of large stone structures crowning elevated mountain-tops, some of them from twenty to thirty feet square. On the Hassayampa, and the mountainous country south from •Prescott, these ruins are numerous, and were evidently built on their commanding positions by people who were constantly harassed by savage foes. That the bed of the Hassayampa has been washed for gold in ages past, is proven by the large pines, whose age is numbered by hundreds of years, found growing where the ancient miner once searched for the precious metal. Prescott, the modern capital of Arizona, occupies, it is believed, the site of an ancient city, and many relics of its former inhabitants which have been brought to light, go to strengthen this theory.

Near Fort McDowell are found the remains of a large fortification, and of an immense irrigating canal. The bones of a man, supposed to be seven feet high, were unearthed near this point. On the Rio Bonito and other branches of the Salt river, numerous cave dwellings are found. The Colorado Chiquito valley exhibits traces of mounds and irrigating ditches, showing that this region was at one time densely populated. All over the Territory, north from the Casa Grande on the Gila, and extending into New Mexico and Southern Colorado, the ruins of stone buildings, large towns, cave dwellings, and immense canals are met with, in the valleys and on the mountain-tops and hillsides, near the principal water-courses.

Nothing is left to tell the story of the people who constructed them, save the few earthen vessels which have been found in the ruins, the stone hammers and axes occasionally met with, and the fragments of broken pottery which lie scattered about their former abode. From the charred remains of human bones taken from the ruins, it has been supposed that the ancient people cremated their dead; and, from the few hieroglyphics which they have left behind, it has been thought they were sun-worshipers. As to their pursuits and mode of life, it is generally believed they followed the business of mining, as well as agriculture. As has been before alluded to, the evidence is conclusive that many of the gulches in the Sierra Prieta range were worked for the golden treasures hundreds of years ago. That this ancient race, who have left such massive monuments of their skill and industry behind them, had made rapid progress in the arts of an advanced civilization, there can be no doubt. Who were those people who erected imposing structures, opened canals, and brought immense stretches of land under cultivation? From whence did they come, and what has been the cause of their extinction, so complete that nothing is left to tell the story? Many theories have been advanced as to their origin and history, but nothing definite is yet known of one of the most remarkable of prehistoric races of the American continent.

Here is a wide field for the savant who desires to trace the evidences of a civilization whose origin is lost in the mists of antiquity, and whose crumbling monuments yet proclaim its ancient vigor and wide extent. Perhaps the key to unlock the barred and bolted chambers of prehistoric American history may yet be found in the ruins of Arizona.

THE EARLY SPANISH MISSIONARIES.

Following closely in the wake of that army of daring adventurers, fired with the thirst for gold and glory, who conquered the vast empire of the Montezumas, and penetrated to the wild regions northward, came another army, which made up in fiery zeal what it lacked in numbers; an army proclaiming " peace on earth and good will to men," whose standard was the emblem of Christianity, and whose mission was the spreading of the gospel among the tribes of the far South-west. Padre de Niza, as has been before stated, was the pioneer of the cross in what is now known as Arizona. He penetrated to the Cities of Cibola, and on his return to New Spain spread glowing reports of their richness and extent, which led to the expedition of Coronado. It has been charitably supposed that the father indulged in this exaggeration in the hope of extending the gospel of Christianity among the natives, but Coronado and his followers, disappointed in not finding the expected treasures, abandoned the country in disgust, and no efforts were made to establish permanent settlements in Arizona until more than a hundred years later.

The first attempt to found missions in this Territory, then known as Pimeria Alta, was made by the Franciscan fathers in 1650, at the Moquis villages. The enterprise was undertaken under the direction of the Duke of Albuquerque, then Viceroy of Mexico. In 1680, the Indians rebelled, massacred many of the Spaniards, and the missions were abandoned and never re-established. As near as can be ascertained, the first mission built in Southern Arizona, was at Guavavi, forty-six miles south of Tucson, in the latter part of the seventeenth century. The mission of Tumacacori was founded some time afterwards, and San Xavier, below Tucson, in 1694. In the same year, Fathers Kino and Mange, who had been active in establishing missions in Sonora, visited the Gila river, and were the first to thoroughly inspect the ruins of the Casa Grande. They also explored the lower Gila and Colorado. Father Kino was a true friend of the Indians, and labored untiringly to better their condition. He procured an order from the Audience of Guadalajara that his neophytes should not be apportioned out to work in the mines. Father Kino established several missions among the Pimas, who made rapid progress in civilization under the parental care of the humane priest. In 1720 there were nine missions in a flourishing condition within the Territory now known as Arizona. They were Tubac, San Xavier del Bac, Joseph de Tumacacori, San Miguel, Guavavi, Calabasas, Arivaca and Santa Ana. They were rich in flocks and herds, and in the products of the silver mines, which they worked extensively.

As showing the mode of life among the converts at the missions, we copy the following, written by Bishop Salpointe of Tucson: " Early in the morning the Indians had to go to church for morning prayers and to hear mass. Breakfast followed this exercise. Soon after, a peculiar ring of the bell called the workmen. They assembled in front of the church, where they were counted by one of the priests, and assigned to the different places where work was to be done. When the priests were in sufficient numbers, they used to superintend the work, laboring themselves, otherwise they employed some trustworthy Mexican to represent them. Towards evening, a little before sundown, the workmen were permitted to go home. On their arrival in the houses, which were located around the plaza, one of the priests, standing in the middle of this plaza, said the evening prayers in a loud voice in the language of the tribe. Every word he pronounced was repeated by some selected Indians, who stood between him and the houses, and last, by all the Indians present in the tribe." Under the fostering care of the fathers, large tracts of land were brought under cultivation, and the Indians appeared to be contented and happy.

In 1744, Father Jacob Sedel made an attempt to reach the Moquis and re-establish the missions, but got no further than the country of the Pimas on the Gila, who dissuaded him from the enterprise. He explored the newly discovered river of Asumpciou (Salado) and the Verde. He also followed the Gila to its sources, and encountered the Apaches. In 1727, the Bishop of Durango, Don Benito Crespo, visited the missions of Arizona, and wrote to Philip V. in their behalf. That monarch ordered that they should be protected and assisted out of the royal treasury. In 1751 there was an outbreak of the Pimas, most of the priests killed, and the missions in the northern part of the province destroyed. The revolt was instigated " by one Luis, from Saric (Sonora), who pretended to be a wizard, and made the Indians consider as a disadvantage to them what he intended' for his own benefit." In 1765, the prosperity of the missions received a heavy blow from the decree ordering the expulsion of the Jesuits from Spain and her colonies. In May, 1768, fourteen Franciscan fathers, from the college of Santa Cruz of Queretaro, arrived at Guaymas, destined to take the place of the Jesuits who had been killed by the Indians, and expelled by order of the government. They found the missions which had escaped the fury of the Indian revolt in a declining condition. Life and energy had fled with the Jesuits; the Apache, till then but little known, had swooped down on the flocks and herds, and the missions seemed to be on the brink of ruin. But, under the unremitting care of the Franciscans, they soon recovered their former flourishing condition.

Captain Bautista Ainsa, under orders from the Viceroy, undertook to open communication by land from Sonora tō Upper California in January, 1774. He was accompanied by Fathers Garcez, Pedro and Ehrarch, who penetrated the country of the Yavapais and explored the central portion of Arizona. Captain Ainsa returned from California in 1776, bringing with him chief Palma and others of the Yuma tribe, praying for the establishment of missions among them. Three missions were estab-

THE EARLY SPANISH MISSIONARIES. 115

lished by Father Garcez in 1779—La Concepcion, where Fort
Yuma now stands, San Pedro, near Castle Dome, and San
Pablo, near Chimney Peak. On the seventeenth of July, 1781, the
Yumas rose in rebellion against the Spanish authorities, killed
the garrison at La Concepcion, and carried the women and
children into captivity. The priests were murdered, the build-
ings destroyed, and thus ended the missions of the Colorado.
No steps were afterwards taken to re-establish them.

Among the adventurous pioneers of the cross who traversed
Arizona from 1773 to 1776, were Fathers Pedro Font, Francisco
Garcia, Silvestre Escalante and Francisco Dominguez. They
explored the Casa Grande ruins and the Moquis villages.
Father Escalante's party went as far north as the Uintah
mountains in Utah, and as far south as Moro, New Mexico.
They crossed the Colorado somewhere in the neighborhood of
latitude 37° north, and between longitude 111° and 112°, west
of Greenwich. Escalante appears to have been the last of the
adventurous missionaries who journeyed through the wilds of
Arizona for nearly two hundred and fifty years subsequent to
the expedition to the Seven Cities. Notwithstanding the raids
of the Apache, the missions of Southern Arizona continued
in a high state of prosperity until the Mexican war of independence. After that they lost the support and protection of
the vice-regal Government, languished and declined, and were
finally suppressed and abandoned by a decree of the Mexican
Government, in 1827.

Of all the mission churches built by the Franciscans and
Jesuits, but one remains in a state of preservation—that of San
Xavier del Bac, nine miles south of Tucson. This, the most
important mission in the Territory, was established in 1694, but
the present building was not commenced until 1768. On the
abandonment of the missions in 1827, the Papago Indians,
who resided at San Xavier, took charge of the church, and
preserved it from destruction by the Apaches. The style of
architecture of San Xavier is a mingling of the Moorish and
the Spanish. It is built of stone and brick, with a fine coating
of cement. It has a length of 105 and a width of 27 feet, inside
the walls. It is in the form of a cross. The nave is divided
into six parts, marked by as many arches. The building is
surmounted by a dome and two towers, one of which remains
unfinished. The church faces to the south, the façade being
ornamented with scroll-work and the coat of arms of the Franciscan order. Around the roof is a brick balustrade, covered with
cement, and with griffins' heads, also in cement, at each angle
and corner. The interior is a mass of elaborate gilding, paint-
ing, and fresco-work. On the right-hand side, between the
front door and the main altar, there is a fresco representing
the "Coming of the Holy Ghost," and on the left, a picture of
the "Last Supper." The main altar is dedicated to St. Francis
Xavier. The frescoes near the altar are the "Adoration of the
Wise Men," the "Flight into Egypt," the "Adoration of the
Shepherds," and the "Annunciation," still in a good state of

preservation. The main altar, and those on either side, are decorated with columns and arabesques in relief, gilded and painted in many colors in the Moorish style. Statues of the twelve apostles are placed in niches in the pillars of the church. The ceilings were adorned with fresco-work, but much of it has been defaced by the rain trickling through the roof.

Near the front door are two small openings communicating with the towers; from these rooms commence the stairs, cut into the thickness of the walls. The second flight brings the visitor to the choir of the church. There are some fine frescoes here. Two flights more lead to the belfry, where hang four home-made bells of small size. Twenty-two steps more lead to the little dome, covering the tower, about seventy-five feet above the ground. From this point a fine view can be had of the beautiful Santa Cruz valley, and the peaks and mountain ranges which surround it in every direction. On the west side of the church is an inclosure and a small chapel. This was formerly used as a cemetery, the bodies being kept in the chapel until the ceremony of burial was performed.

When we remember the age in which it was built, and the facilities at hand for its construction, the church of San Xavier must be considered a remarkable structure. The traveler who first beholds its perfect outlines, standing in solitary grandeur on the edge of the desert plain, is astonished to find in this remote region a building which would adorn any capital in Christendom. It stands an impressive monument to the untiring zeal, energy, and self-sacrificing devotion of the mission fathers, who penetrated the unknown wilds of the south-west, and were the first to open to settlement and civilization, what is now the Territory of Arizona. The effects of their early labors are yet seen in the tribes they redeemed from barbarism and taught the arts of civilization, peace, and industry. The only other relic of the missions found in the Territory is the ruins of St. Joseph at Tumacacori, three miles below Tubac, on the Santa Cruz river. This mission was destroyed by the Apaches in 1820, and the occupants massacred. The building was smaller and of ruder construction than San Xavier. The form was that of a Greek cross with a basilica. The latter is still standing, crowned by the emblem of Christianity. Two towers yet remain in a fair state of preservation. The church was built of adobe, plastered with cement, and coped with burnt brick. The roof was flat and covered with tiles. The valley adjacent to this mission was brought under a high state of cultivation. Tumacacori was at one time the richest of the Arizona missions, and was the scene of an active and prosperous mining industry, but the Apache spoiler "came down like a wolf on the fold," and nothing remains to tell of Jesuit energy and endeavor, save the crumbling ruin of the old church and the abandoned shafts and tunnels, overgrown with brush and filled with debris, which are frequently met with in the surrounding mountains.

HOW TO GET TO ARIZONA.

To reach Southern Arizona from the East, at the present time, the shortest and most direct route is by way of the Atchison, Topeka, and Santa Fe railroad. This line begins at Kansas City, Missouri, and, passing through Kansas, Colorado, and New Mexico, unites with the Southern Pacific at Deming, 1,149 miles from Kansas City; fare $74, first class. From Deming to Benson, twenty-eight miles from Tombstone, it is 173 miles; fare, $17 30.' Daily stage lines run from Benson to Tombstone; fare, $6. From Deming to Tucson it is 219 miles; fare, $21 90—thus making the distance from Kansas City to Tombstone 1,340 miles, and to Tucson, 1,368 miles. Sleeping-cars are run on this route, and passengers have every comfort found in railroad traveling. The time from Kansas City to Tombstone or Tucson is about three days.

To reach Northern Arizona from the East, the traveler takes the Atchison, Topeka, and Santa Fe line to Albuquerque, New Mexico. At this point the Atlantic and Pacific railroad strikes westward, on the thirty-fifth parallel, through Northern Arizona. This road is completed as far as Brigham City, in Apache county, 280 miles from Albuquerque. The fare from Kansas City to Albuquerque is $53. Persons desirous of visiting Northern Arizona will find stages at Brigham City, or at the end of the track, to convey them to Prescott and the principal points in Apache, Yavapai, and Mohave counties. Brigham City is about 180 miles east of Prescott, but the railroad is advancing at the rate of more than a mile a day, and the track will be 50 miles north of the capital of Arizona by the first of July, 1882. Prescott is distant from Kansas City 1,368 miles.

To reach Arizona from California, or the Pacific coast States or Territories, the quickest route is by the Southern Pacific railroad. To North Arizona by this line, the traveler has the choice of two routes from Yuma, by steamer up the Colorado, or by rail to Maricopa. Below we append a table of distances and rates of fare by this route to the principal points in the Territory, from San Francisco:

Aubrey, Mohave county—Southern Pacific railroad to Yuma, 731 miles; river steamer, 255 miles; fare, $65.

Benson, Cachise county—Southern Pacific railroad, 1,024 miles; fare, $58.

Casa Grande, Pinal county—Southern Pacific railroad, 913 miles; fare, $52.

Castle Dome, Yuma county—Southern Pacific railroad, to Yuma, 731 miles; river steamer, 22 miles; fare, $49.

Florence, Pinal county—Southern Pacific railroad, to Casa Grande, 913 miles; stage, 22 miles; fare, $57.

Globe City, Gila county—Southern Pacific railroad, to Casa Grande, 913 miles; stage, via Florence; fare, $72.

Mineral Park, Mohave county—Southern Pacific railroad, to Yuma, 731 miles; river steamer to Hardyville, 300 miles; stage, 43 miles; fare, $75.

Pantano (station for Harshaw), Pima county—Southern Pacific railroad, 1,006 miles; fare, $57; by stage to Harshaw, 50 miles.

Phœnix, Maricopa county—Southern Pacific railroad, to Maricopa, 887 miles; stage, 35 miles; fare, $55.

Prescott, Yavapai county—Southern Pacific railroad, to Maricopa, 887 miles; stage, 150 miles; fare, $75.

Tombstone, Cachise county—Southern Pacific railroad, to Benson, 1,024 miles; stage, 31 miles; fare, $62.

Tucson, Pima county—Southern Pacific railroad, 978 miles; fare, $55.

Wilcox, Cachise county—Southern Pacific railroad, 1,064 miles; fare, $60.

The fares quoted above are first class. The local rate charged by the Southern Pacific in Arizona is ten cents per mile. From the foregoing it will be seen that all the principal points in Arizona can be visited from the East or the West quickly and comfortably; giving the traveler choice of rail, river, and stage routes through the Territory.

THE WANTS OF THE TERRITORY.

In the foregoing pages has been given a " brief chronicle " of the Territory, its past history, its present condition, and its future prospects. Before closing this short sketch of the country and its resources, it may not be out of place to note the aids which it needs to bear it on to the topmost wave of material prosperity. Arizona wants, first of all, capital to develop her vast mineral wealth; she wants men who have the enterprise and the means to open up the treasures which lie hidden in her mountains and *mesas*, to sink shafts, to drive tunnels, to erect mills and furnaces, to give employment to labor, to build up happy homes and thriving communities, and send forth such a volume of bullion as has never been equaled in the history of the globe. As mining is the leading industry of the country, the capital to place that industry on a prosperous basis is a vital necessity for the welfare of Arizona. Here are gold, silver, copper, coal, lead, and iron scattered in profusion throughout the length and breadth of the Territory; here are railroads penetrating in every direction; here is a climate of almost perennial summer, and here is every natural facility for the extraction and reduction of ores. For the men who are waiting in the East and in Europe for a chance to invest some of their surplus millions, here is a land with grand resources almost untouched, offering opportunities for profitable mining ventures not equaled in the western country, and only awaiting the magic wand of capital to cause its mountains and hills to send forth streams of treasure.

As has been remarked in another place, Arizona wants men

who will engage in manufacturing enterprises. Hundreds or thousands of dollars are annually sent out of the country for supplies which could be produced at home. The manufacture of woolen goods, of leather, of soap and candles, and many other articles, offers almost certain assurance of success. For the man or men with a knowledge of the business and the requisite capital, who will engage in any of these enterprises, a fortune is in store.

There are yet millions of acres of unoccupied grazing land in the Territory, waiting for the cattle raiser to utilize its fine grasses. On portions of this immense domain water is scarce, but the want can be quickly supplied by the sinking of wells. No finer climate for stock can be found, and no better beef is raised in the United States. There is plenty of room for twice the number of cattle now in the Territory, and with two railroads crossing it from east to west, and leading to the markets of the Atlantic and the Pacific, no better field for this branch of industry can be found.

To men who have some means, and can take advantage of the opportunities that present themselves in a new country, Arizona offers an inviting field for the display of their industry, energy, and enterprise. For live, active men, with plenty of "push" and vim, there is always an opening. Arizona wants men with strong hands and stout hearts; men who are willing to work; men who are not afraid to rough it in a new country; men who can fight the battle of life, and are not disposed to give up the contest because fortune does not always smile on them; men who are not above turning their hands to anything that presents itself; men who are sober, steady, and industrious. With such a class of men, to build up the country and develop its grand resources, Arizona will soon become one of the foremost States in the American Union.

We have briefly stated here the character of the emigration which the Territory wishes to attract within its borders; it may be in order, also, to allude to the kind it doesn't want. Of lawyers and doctors the Territory has more than enough, and an influx of the "learned professions" is not desirable. They are already overcrowded, and sharp competition has made the practice of law and medicine anything but profitable. It is true, in these, as in all other professions, "there is room at the top," but unless a man has the acquirements and the talents to take that position, he had better remain where he is. Of clerks, and all those who are seeking desirable positions, where the labor is light and the salary high, the supply on hand already exceeds the demand, and such persons had better stay where they are, unless they are willing to take hold of anything that presents itself, from driving a bull-team to "polishing the head of a drill."

That large class who imagine their fortunes would be made if they could only get to the West, without scarcely an effort on their part, need not come to Arizona. No drones in the hive of industry are wanted here. As everywhere else, energy, perseverance, and hard work, are required for success, and he who thinks to achieve it by any other means will be sadly disappointed, and should remain " at home at ease." Of that grand army of fault finders, never satisfied and forever complaining, this Territory wants none; men who sit supinely waiting for fortune to bid them good-morrow, who make no effort to help themselves, and then complain of their non-success, should not come to Arizona.

In this short space we have alluded to the class of emigration which this Territory is in need of, and also that class it can well afford to do without. There is here plenty of room for an active, enterprising, energetic class of people; who will open our mines, cover our plains and hillsides with flocks and herds, cultivate our rich valleys, build up happy homes and prosperous communities, and by industry, enterprise, temperance, and integrity lay broad and deep the foundations of the coming great State of the South-west.

www.ingramcontent.com/pod-product-compliance
Lightning Source LLC
Chambersburg PA
CBHW021939190326
41519CB00009B/1077